Dilute Magnetic Semiconducting (DMS) Materials

by R. Saravanan

Diluted Magnetic Semiconductors (DMS) play a vital role in modern electronics industry. It is important to understand the fundamental properties of these materials in order to apply them to their full potential.

This book presents an analysis of the charge density distribution and other properties of some silicon and germanium based diluted magnetic semiconductors. A quantitative analysis of the charge density distribution has been done in order to obtain measurements of the charges involved in the bonding, which are decisive for the physical and chemical properties of the DMS materials. Also, the local structures of the materials have been analyzed by studying their powder X-ray diffraction intensities. Analysis of the magnetic properties of the DMS materials is mandatory and has been accomplished by magnetic measurements carried out using a vibrating sample magnetometer. The morphology of the DMS materials has been studied using scanning electron micrographs.

Keywords: Diluted Magnetic Semiconductors (DMS), Magnetic Semiconductors, Vanadium Doped Germanium, Manganese Doped Germanium, Cobalt Doped Germanium, Manganese Doped Silicon, Nickel Doped Silicon

Dilute Magnetic Semiconducting (DMS) Materials

by

Dr. R. Saravanan, M.Sc., M.Phil., Ph.D.

Associate Professor & Head

Research Centre and PG Department of Physics

The Madura College (Autonomous)

Madurai - 625 011

India

Published by **Materials Research Forum LLC**
Millersville, PA 17551, USA

Published as part of the book series
Materials Research Foundations
Volume 35 (2018)
ISSN 2471-8890 (Print)
ISSN 2471-8904 (Online)

Print ISBN 978-1-945291-76-0
ePDF ISBN 978-1-945291-77-7

Distributed worldwide by

Materials Research Forum LLC
105 Springdale Lane
Millersville, PA 17551
USA
http://www.mrforum.com

Manufactured in the United States of America
10 9 8 7 6 5 4 3 2 1

Table of Contents

Preface

Diluted magnetic semiconductors (DMS) find a vital role in the modern electronics industry. It is important to understand the properties of the materials in order to apply them to modern electronics. Various methods are adopted for the fabrication and characterization of these materials. Melt growth technique, thin film deposition, mechanical alloying and arc melting technique are some of the important methods used for the fabrication of DMS materials. Analysis on structural properties, electrical properties and magnetic properties are carried out to understand the characteristics of the fabricated DMS materials.

Among the available characterization techniques, the powder X-ray diffraction technique finds an inevitable position, since it provides finite details of the materials in the form of charge density information. In this book, a quantitative analysis of the charge density has been done for the materials in order to obtain measurements of the charges involved in the bonding, which decide the physical and chemical properties of the prepared DMS material. Also, the local structure of the material has been analyzed using the observed powder X-ray diffraction intensities. Analysis of the magnetic properties of the DMS materials is mandatory and has been accomplished by the magnetic measurements carried out using a vibrating sample magnetometer. The morphology of the DMS materials has been studied using the respective scanning electron micrographs. This book elaborates the analysis of the charge density distribution and other properties of some silicon and germanium based diluted magnetic semiconductors.

Chapter 1 gives an introduction of various types of semiconductors, diluted magnetic semiconductors and their applications. It also gives a survey on some earlier research works done on diluted magnetic semiconductors. Information on the methods of preparation of DMS materials is given in this chapter as well. Also, characterization techniques such as powder X-ray diffraction technique, scanning electron microscopy (SEM), energy dispersive X-ray analysis (EDAX) and magnetic hysteresis measurements using vibrating sample magnetometer (VSM) are explained in this chapter.

Chapter 2 introduces the analytical techniques used to analyze the charge density of the proposed DMS materials. Analysis based on the obtained powder X-ray diffraction intensities is explained in this chapter. The detailed theory of charge density analysis has been presented along with the mathematical tools used for them. An introduction on the analysis of the structural properties and electronic charge density distribution of materials is given in this chapter. An analysis of the local structure of the materials is also elaborated. A review on some earlier works done on the charge density analysis of materials is also elaborated here.

Chapter 3 deals with the method of preparation of the samples of the proposed diluted magnetic semiconductor materials and the structural analysis of the prepared DMS materials.

The materials prepared using the high temperature melt growth technique are as follows;

$Ge_{1-x}Mn_x$ (x = 0, 0.04, 0.06, 0.10); $Ge_{1-x}V_x$ (x = 0.03, 0.06, 0.09); $Ge_{1-x}Co_x$ (x = 0.03, 0.06, 0.09).

The materials prepared using the ball milling technique are as follows;

$Si_{1-x}Mn_x$ (x = 0.02); sample milled for 100h and 200h

$Si_{1-x}Ni_x$ (x = 0, 0.03, 0.06, 0.09, 0.12).

The morphology of the proposed materials is studied and presented in this chapter. The structural analysis of the proposed materials is done and the results are presented in this chapter.

Chapter 4 gives a detailed analysis of the theoretical charge density distribution of the proposed materials. This theoretical estimation of the charge density of the materials is useful to predict the properties of the DMS materials before moving on to the experimental understanding. The charge density of the Ge and Si based DMS materials with the dopants of manganese (Mn), cobalt (Co) and vanadium (V) is estimated theoretically and analyzed using the charge density maps. The results are presented in this chapter.

Chapter 5 deals with the estimation and the analysis of the experimental charge density of the prepared materials which are derived from the structure factors obtained from the observed powder X-ray diffraction intensities. The quantitative analysis of the estimated experimental charge density distribution of the grown DMS materials is done using the mapping of charge density in 3-dimensional, 2-dimensional and 1-dimensional spaces.

Chapter 6 gives the results of the analysis of the magnetic measurements of the prepared samples carried out using vibrating sample magnetometer. This chapter also gives the correlation of the observed magnetic properties to the structural properties and the distribution of charge density of the prepared DMS materials.

Chapter 7 deals with the analysis of local structure of the prepared DMS materials using the observed powder X-ray diffraction intensities. This analysis is done using the pair distribution function of the materials. The distances of the nearest neighbors are tabulated, which are helpful in quantifying any deviation or strain in the lattice, which can possibly occur due to the process of doping.

Chapter 8 lists down the consolidated discussion of this work.

Parts of the results of this book have been published as follows:

1.   Magnetism in melt grown dilute magnetic semiconductor $Ge_{1-x}Mn_x$ from electron density, R. A. J. R. Sheeba, R. Saravanan and L. John Berchmans, Materials Science in Semiconductor Processing, 15, 731 – 739 (2012)

2.   Magnetic and charge derived properties of ball milled dilute magnetic semiconductor $Si_{0.98}Mn_{0.02}$, *R. A. J. R. Sheeba*, R. Saravanan and S. Sasikumar, *Physica B*, 426, 71 – 78 (2013)

3.   Signature of antiferromagnetism in entropy maximized charge density distribution of melt   grown diluted magnetic semiconductor, $Ge_{1-x}V_x$, *R. A. J. R. Sheeba*, R. Saravanan and L. John Berchmans, *Journal of Materials Science: Materials in Electronics*, 26 No.6, 3772 – 3780 (2015)

4.   Understanding electronic and magnetic transitions in ball milled diluted magnetic semiconductor $Si_{1-x}Ni_x$ through experimental electron density distribution, *R. A. J. R. Sheeba*, S. Saravanakumar, S. Israel and R. Saravanan, *Journal of Alloys and Compounds*, 728, 887 – 895 (2017)

Dilute Magnetic Semiconducting (DMS) Materials, R. Saravanan   Materials Research Forum LLC
Materials Research Foundations **35** (2018)   doi: http://dx.doi.org/10.21741/9781945291777

Chapter 1

Introduction

Abstract

Chapter 1 gives an introduction of various types of semiconductors, diluted magnetic semiconductors and their applications. It also gives a survey on some earlier research works done on diluted magnetic semiconductors. Information on the methods of preparation of DMS materials is given in this chapter as well. Characterization techniques such as powder X-ray diffraction technique, scanning electron microscopy (SEM), energy dispersive X-ray analysis (EDAX) and magnetic hysteresis measurements using vibrating sample magnetometer (VSM) are explained in this chapter.

Keywords

DMS, Magnetic Semiconductors, Silicon, Germanium, X-ray, EDAX, Magnetometer

Contents

1.1　Objectives

The main objective of this book is to report the preparation of silicon and germanium based diluted magnetic semiconductors (DMS) and to report the results of the characterization of the prepared materials for their structural properties, distribution of charges, magnetic properties and morphological nature.

Diluted magnetic semiconductor (DMS) materials have been extensively investigated due to their enormous applications in the field of electronics such as spintronics, photonics *etc.* The present thesis analyzes silicon (Si) and germanium (Ge) based DMS materials because of their potential applications in spintronics devices such as (i) magneto resistive random access memory (MRAM) (ii) spin transistor (iii) giant magneto resistive (GMR) read heads for hard disks (iv) magnetic sensors *etc.*

The silicon (Si) and germanium (Ge) based diluted magnetic semiconductor (DMS) materials considered for the present investigation are:

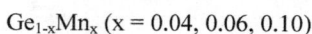

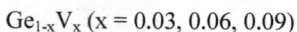

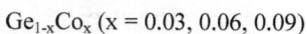

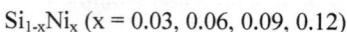

$Ge_{1-x}Mn_x$ (x = 0.04, 0.06, 0.10)

$Ge_{1-x}V_x$ (x = 0.03, 0.06, 0.09)

$Ge_{1-x}Co_x$ (x = 0.03, 0.06, 0.09)

$Si_{1-x}Mn_x$ (x = 0.02)

$Si_{1-x}Ni_x$ (x = 0.03, 0.06, 0.09, 0.12)

The present work reported in this book has the following tasks.

1.　　Growth of Ge based diluted magnetic semiconductors, $Ge_{1-x}Mn_x$, $Ge_{1-x}V_x$ and $Ge_{1-x}Co_x$, using high temperature melt growth technique.

2.　　Prepartion of Si based diluted magnetic semiconductors, $Si_{1-x}Mn_x$ and $Si_{1-x}Ni_x$ using ball milling technique.

3. Analysis of the structural properties of the grown DMS samples by powder X-ray diffraction (PXRD) technique employing Rietveld (Rietveld, 1969) method. The execution of the Rietveld (Rietveld, 1969) analysis is carried out using the software package JANA 2006 (Petříček et al., 2014).

4. Analysis of the electronic charge density distribution between atoms in the lattice planes of the prepared DMS materials using maximum entropy method (MEM) (Collins, 1982). The determination of the electronic charge densities in the grown DMS materials has been carried out using the software PRIMA (Izumi et al., 2002) and the charge density distribution in the unit cell has been visualized using the software VESTA (Momma et al., 2011). This study has been carried out for the first time for the DMS materials.

5. Analysis of the surface morphology and microstructure through the micrographs recorded using scanning electron microscope (SEM).

6. Analysis of the local structure of the grown DMS materials using pair distribution function (PDF) (Proffen et al., 1999). The pair distribution function of the grown DMS samples has been obtained through the software PDFgetX (Jeong et al., 2001). The observed pair distribution functions of the DMS samples were fitted using PDFgui (Farrow et al., 2007).

7. Analysis of the magnetic properties of the prepared Si and Ge based DMS materials using the magnetic hysteresis recorded using vibrating sample magnetometer (VSM).

The details of the preparation of the DMS samples, the results obtained from various characterization techniques and the conclusion drawn from the results are given in the following chapters.

1.2 Introduction of semiconductors

Semiconductors are materials, whose electrical properties lie between conductors and insulators. The valence band and conduction band overlap in conductors, thus enabling free motion of the charge carriers inside the material, without any application of external energy. In the case of insulators, the valence and conduction bands are separated by an energy gap, which is large so that the charges from valence band cannot move to the conduction band. In semiconductors, the valence and the conduction bands are separated by an energy gap lower than that of insulators. By the application of external energy, the charge carriers in the valence band can be made to move to the conduction band, which enhances the conductivity of the material. The classification of materials is shown in figure 1.1.

Semiconductors can be classified into two major groups. They are;

I. Elemental semiconductors and

II. Compound semiconductors

1.2.1 Elemental semiconductors

Semiconductors comprising a single element are called elemental semiconductors. They normally crystallize in diamond structure and hence, the nature of bonding between the atoms is covalent; example: Silicon (Si) and Germanium (Ge). Silicon (Si) and germanium (Ge) are elemental semiconductors, which are used vastly in the electronics industry. Silicon is used because of its natural abundance in the form of SiO_2, which can be purified to obtain silicon. Si forms the basis for the semiconductor devices such as transistors, diodes *etc.*, which are the foundation of modern electronic devices like radio, computers, telephones *etc.* The response of Si-based devices is fast and they are far more resistant to heat damage because of the higher melting point of Si (1414°C) (Haynes, 2014). Due to the heat resistant property of Si, it remains a semiconductor even at high temperatures. Germanium can be substituted in place of silicon as it is compatible with Si-based technology (Kittel, 1976). Hence, germanium based semiconductor devices also play a vital role in the modern electronics industry.

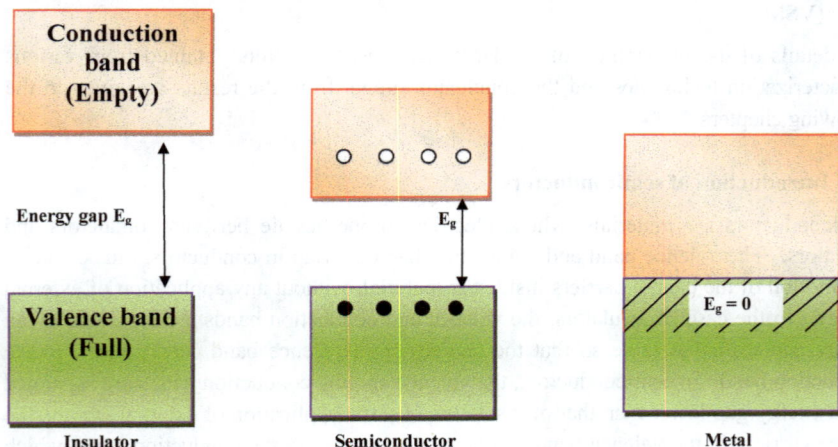

Figure 1.1 Comparison of semiconductor, metal and insulator.

1.2.2 Compound semiconductors

Semiconductors composed of two or more different elements are called compound semiconductors. Elements combined from groups II and VI of periodic table are called II–VI semiconductors and elements combined from groups III and V of periodic table are called III–V semiconductors (Kittel, 1976). Some of the examples are given below:

a) II–VI semiconductors

1. Cadmium Selenide (CdSe)

2. Cadmium Telluride (CdTe)

3. Zinc Sulphide (ZnS)

b) III–V semiconductors:

1. Gallium Arsenide (GaAs)

2. Gallium Phosphide (GaP)

3. Gallium Nitride (GaN)

4. Indium Phosphide (InP)

The compound semiconductors have some advantages over elemental semiconductors. The main advantage is that, the mobility of the charge carriers is higher in compound semiconductors. This higher mobility enhances the speed of operation of the semiconductor devices. Also, the energy to be supplied to initiate conduction, *i.e.*, the threshold is lower than that of the elemental semiconductors. Some compound semiconductors (*eg.* GaAs) have wider band gap, which can be used in high temperature power operations with lower thermal noise (Milton Ohring, 1998).

Some of the compound semiconductors are used in opto-electronic applications like semiconductor lasers, light emitting diodes, *etc.* (*eg.* GaAs, In doped GaP, InP, *etc.*)

Semiconductors can be further classified into two categories:

a) Intrinsic semiconductors

b) Extrinsic semiconductors

1.2.3 Intrinsic semiconductors

Semiconductors in the pure form are called "intrinsic" semiconductors. Examples of intrinsic semiconductors are silicon (Si) and germanium (Ge). The intrinsic semiconductors have limited applications, because of the limited number of charge

Dilute Magnetic Semiconducting (DMS) Materials, R. Saravanan   Materials Research Forum LLC
Materials Research Foundations **35** (2018)   doi: http://dx.doi.org/10.21741/9781945291777

carriers in the material, and the conductivity of the material mainly depends on temperature (Anthony, 1987).

1.2.4 Extrinsic semiconductors

The conductivity of a semiconductor can be improved by decreasing its resistance by altering the temperature, since the value of the resistance in a semiconductor decreases with increase in temperature. The conducting properties of semiconductors can also be modified by deliberate addition of impurities in a controlled fashion in the semiconductor crystal. The added impurity element is called "dopant", and the process of adding the dopant is called "doping". The doped semiconductor is called "extrinsic semiconductor". The process of doping lowers the resistance of the doped semiconducting material, because of the increase in the number of the charge carriers. Also, semiconductor junctions are created between differently doped regions of extrinsic semiconductors and these regions are called depletion regions. An extrinsic semiconductor having free holes as majority carriers is called a p-type semiconductor, while, an n-type semiconductor has free electrons as majority carriers. In order to fabricate the electronic devices, the semiconductors are doped under precise conditions, to control the concentration and regions of p- and n-type dopants. The p-n junctions between these regions are responsible for the useful electronic behavior.

Based on the electronic properties, the semiconductors can be further classified into two types:

(i) Direct band gap semiconductors

(ii) Indirect band gap semiconductors

The minimum energy state in the conduction band and the maximum energy state in the valence band are characterized by the crystal momentum, $\vec{k}$ in the Brillouin zone. If the valence and the conduction bands have the same momentum, then, it is known as the direct band gap semiconductor (Figure 1.2). In this case, the electron can directly emit a photon. If the momentum is varied for the valence and the conduction bands in the material, then, it is called an indirect band gap semiconductor (Figure 1.3). In indirect band gap semiconductor, the electron cannot directly emit a photon since, it has to go through an intermediate stage in which, a phonon is emitted and the momentum is transferred to the crystal lattice. Some examples of direct and indirect band gap semiconductors are given below as table 1.1.

Dilute Magnetic Semiconducting (DMS) Materials, R. Saravanan   Materials Research Forum LLC
Materials Research Foundations **35** (2018)   doi: http://dx.doi.org/10.21741/9781945291777

Table 1.1 *Examples of direct band gap and indirect band gap semiconductors*

Direct band gap semiconductors	Indirect band gap semiconductors
Aluminium nitride (AlN)	Diamond (C)
Gallium nitride (GaN)	Silicon (Si)
Indium phosphide (InP)	Germanium (Ge)
Indium nitride (InN)	Gallium selenide (GaSe)
Cadmium selenide (CdSe)	Gallium phosphide (GaP)
Cadmium telluride (CdTe)	Aluminium antimonide (AlSb)

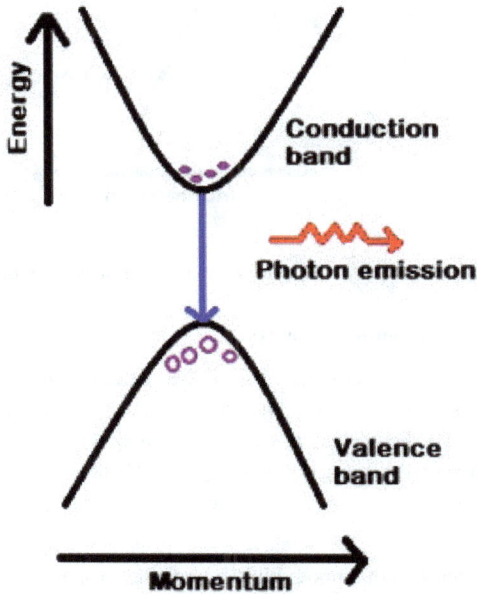

Figure 1.2 Schematic representation of direct band gap of a semiconductor.

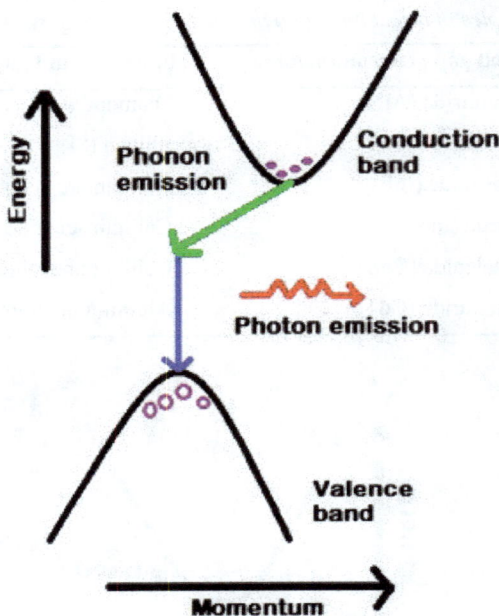

Figure 1.3 Schematic representation of indirect band gap of a semiconductor.

1.3   Diluted magnetic semiconductors

Diluted magnetic semiconductors (DMS), a class of materials, are important to the field of spintronics (Wilamowski, 2006). In semiconductors, substitution of a transition metal element in place of a host element, adds local magnetic moments to the low-energy degrees of freedom of the semiconductor system (Peaton et al., 2003). The transition element is usually substituted on a small fraction (say x), at a host semiconductor element site.

Diluted magnetic semiconductors exhibit interesting properties such as, extremely large Zeeman splitting of electronic bands (Jung Ho Yu et al., 2010) and giant Faraday rotation (Savchuk et al., 2011). Magnetic properties of DMS materials such as, spin glass behavior due to lattice frustration, magnon excitations in random system of spins, super exchange and short range antiferromagnetic (AFM) order are of considerable interest (Furdyna, 1987). DMS materials are found to exhibit optical properties also. Optical conductivity has been observed in DMS material $In_{1-x}Mn_xAs$ that enables it to be used in optoelectronic applications (Hirakawa et al., 2001). The association between the optical, electronic and magnetic properties of Mn–doped III–V semiconductors is of much interest, because of their potential application in future spin-optoelectronic devices (Burch et al., 2008).

The properties of DMS materials can be tuned according to the requirements, by varying the concentration of the magnetic ions doped in the host semiconductor. Band gaps and lattice constants of DMS materials can be tuned by varying the dopant concentration, which in turn modify the properties of DMS materials. The tunability of the lattice parameters provides the advantages of growing quantum wells and superlattices. The large Zeeman effect allows us to tune the depth of quantum wells using an external magnetic field (Lee et al., 1996).

1.3.1 Applications of diluted magnetic semiconductors

1.3.1. (a) Spintronics

Diluted magnetic semiconductors find an imperative role in the field of spintronics. Spintronics stands for spin transition electronics. As we know, electronic circuits in integrated circuit chips used for information processing and communication, employ controlling the charge of electrons in semiconductors, whereas, mass storage of information is carried out by magnetic recording using electron spins in ferromagnetic materials. In spintronics, both the charge and spin of electrons are used at the same time to enhance the performance of devices. The realization of the spintronic devices requires materials with ferromagnetic ordering at operational temperatures compatible with existing semiconductor materials. Since diluted magnetic semiconductors (DMS) are semiconductors with favorable magnetic properties, these materials will suit the requirements for spintronic applications.

Spintronic materials can be employed for the fabrication of storage devices such as Magnetic Random Access Memory (MRAM) chips. MRAM is advantageous over Static RAM (SRAM) and Dynamic RAM (DRAM) due to non-volatile memory (Johnson et al., 1998), very high speed of operation and low power consumption (Pohm et al., 1990). MRAM is capable of becoming a universal solid state memory replacing all the other memory devices available today. If MRAM is used, there will be no booting from a disk drive, stand-by power will be at a minimum and ultimately the computer system can be integrated on a single chip (Akerman, 2005).

Spin transistor is a magnetically sensitive transistor (Datta, 1990), that employs spin up and spin down states of electrons for its operation and storage of information. The main advantage of the spin transistor over the existing semiconductor transistor device is that,

Dilute Magnetic Semiconducting (DMS) Materials, R. Saravanan   Materials Research Forum LLC
Materials Research Foundations **35** (2018)   doi: http://dx.doi.org/10.21741/9781945291777

the spin states of the electrons can be detected and altered without applying an electric current. This results in the reduction of size of the hardware devices and increase in sensitivity. The spin transistor can be used to create cost effective, non-volatile solid state storage devices. The schematic representation of a spin transistor is shown in figure 1.4.

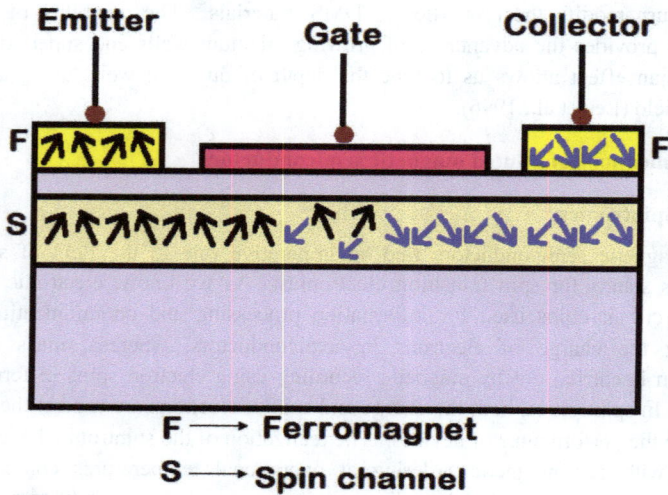

Figure 1.4 Schematic representation of a spin transistor.

Spintronic devices have many advantages such as; smaller size, more versatile and robust than those currently used silicon chips and circuit elements. The information is stored into spins of electrons (up or down). The spins being attached to mobile electrons, carry the information along a wire and the information is read at a terminal using read heads. The spin enables them to retain information even when electrical power is turned off whereas the charge-based data disappears. The spin orientations of conduction electrons survive for a relatively long time due to which, spintronic devices are used particularly for memory storage and magnetic sensors application. Spin interaction is small compared to Coulomb interaction, resulting in less interference. The spin current can flow with less heat dissipation. Miniaturization of devices can be achieved by using spintronic devices.

1.3.1. (b) Photonics

Photonics is a branch of science which deals with generation, detection and manipulation of light through emission, transmission, modulation, signal processing, switching, amplification and detection. The wavelengths used in photonics are mainly in the visible and near infra-red region. Classical optics involves refracting lenses, reflecting mirrors and other optical components and instruments. It does not depend on the quantum properties of light. But in modern optics, photonics is related to quantum optics, optoelectronics or electro optics. Optoelectronics involves devices that comprise both electrical and optical functions (Okabayashi et al., 1998).

Photonics has a wide variety of applications at all levels, *viz.*, devices we use in our domestic life to devices used in technological advancement. For example, light detection, telecommunication, information processing, laser material processing, biophotonics, agriculture, robotics, *etc.* Semiconductor photonic devices include optical data recording, fiber optic telecommunication, laser printing, displays and optical pumping of high-power lasers. Semiconductor light sources like light-emitting diodes, super luminescent diodes, lasers, single photon sources, fluorescent lamps and plasma screens are some examples of photonic applications. The III-V semiconductors are normally used for the applications involving light sources (Burch et al., 2008). Photonic crystal fibers are used in transmission media. Semiconductor optical amplifiers are used to amplify an optical signal. Photo detectors like photodiodes are used to detect light. Modulation of a light source can be achieved by using an optical modulator. Photonic integrated circuits are very useful in optical communication purposes, through their ability to increase the available bandwidth and performance, without significant cost increase. Indium phosphide systems are commonly used in photonic integrated circuits (Vittotio et al., 2012). The photonic devices can create a revolution in optical communication applications.

1.4   Diluted magnetic semiconductors - A review

In the past two decades, the study of magnetic impurities in semiconductors has established a huge consideration. The interest in these diluted magnetic semiconductors (DMS) is motivated by the aspiration to exploit the spin degree of freedom in electronics. The rapidly expanding field of spin based electronics *i.e.*, spintronics, has utilized metal based devices to create a revolution in magnetic sensors and computer hard drive density (Grochowski et al., 2003).

The discovery of ferromagnetism in III-V semiconductor hosts doped with Mn has gained much interest in magnetic semiconductors research since III-Mn-V systems offer a high promise to spintronic applications.

The Mn doped III-V compound semiconductors, (Ga,Mn)As, prepared by molecular beam epitaxial growth revealed ferromagnetic behavior (Ohno et al., 1998). Magneto transport measurements revealed that the critical temperature (T_c) of the grown (Ga,Mn)As sample was 110K. It was found that, the magnetic coupling between two ferromagnetic (FM) (Ga,Mn)As films separated by a non magnetic layer, signified the function of holes in the magnetic coupling. The magnetic coupling between the two FM layers of (Ga,Mn)As, along with the possibility of spin filtering in the resistance temperature detectors (RTDs), show the potential of this particular DMS material for developing new functionality towards future electronics (Wilamowski et al., 2006).

Munekata et al., (1989) fabricated InAs:Mn DMS films using molecular beam epitaxy at two temperatures 200°C and 300°C. The InAs:Mn film grown at 200°C exhibited paramagnetism, whereas, the film grown at 300°C exhibited ferromagnetic property. As the concentration of the dopant Mn was increased, a decrease in lattice constant was observed in the grown films. The band gaps were found to be reduced in both the fabricated films of InAs:Mn (Munekata et al., 1989).

Transition metal (TM) doped II-VI compound semiconductors (Furdyna, 1988) have been extensively studied where TM is primarily the $3d$ transition metals. The close relationship between the sp band electrons of the DMS materials with the band structure of the non magnetic host semiconductors and the characteristics of the nearly localized narrow band arising from the half filled Mn $3d^5$ shells are established in this work. The effect of these on the optical properties of the DMS materials is also elaborated. The mechanism of Mn^+ and Mn^{2+} exchange which is the cause of the magnetism in the DMS materials is proposed. The sp-d exchange interaction between the sp band electrons of the DMS and the $3d^5$ electrons associated with the Mn atom is analyzed followed by the physical consequences such as giant Faraday rotation, magnetic field-induced metal-insulator transition in the chosen DMS. The application of the exchange interaction in quantum wells and superlattices is also described in this work.

Group IV elements are chosen widely for the fabrication of DMS materials, (mainly Ge doped with transition metal ions) due to their compatibility with Si-based technology (Park et al., 2002). In a work by Cho et al. (2002), Mn-doped bulk Ge single crystals were fabricated and studied. The dopant was incorporated in the host lattice up to a concentration of 6%. It was inferred that the lattice constant increases with the concentration of the dopant due to the larger atomic radius of Mn than Ge, which indicates that the dopant Mn atoms were duly incorporated into the Ge host lattice. Lower concentrations of Mn exhibited paramagnetic behavior due to localized magnetic ions while the sample with 6% Mn concentration exhibited ferromagnetic ordering at 285K (Cho et al., 2002).

In a work done by Li et al. (2005), the molecular beam epitaxial method has been adopted to grow Mn-doped Ge and analysis has been done on magnetism and magneto transport properties. Ferromagnetism has been observed in the prepared $Ge_{1-x}Mn_x$ that comes from the clustered dopants associated with inhomogeneities. The anomalous Hall effect occurring between the two critical temperatures was found to be influenced by the induced magnetoresistance. The resistivity measurements of $Ge_{1-x}Mn_x$ showed a metal-insulator transition down to 18 K (Li et al., 2005).

Long range and temperature dependent ferromagnetic behavior has been observed (Morresi et al., 2006) in $Ge_{1-x}Mn_x$ thin films grown by molecular beam epitaxy. It was observed that, isolated nanoclusters of Mn_5Ge_3 about 100 nm in size formed on the surface of the film influenced the magnetic properties of the film. The analysis of the electronic properties indicated that, the presence of substitutional Mn ions dispersed in the host Ge lattice contributes to the formation of the diluted phase $Ge_{1-x}Mn_x$. The contribution of the additional phase Mn_5Ge_3 was negligible because of its very low density. The electrical behavior of the $Ge_{1-x}Mn_x$ thin films indicated a saturation effect with increase in Mn concentration in the film above a dopant concentration of x = 0.03. The semiconductor like behavior and the p-type conduction were due to the residual Mn atoms dispersed in the Ge lattice.

DMS layers of $Ge_{1-x}Mn_x$ (Ayoub et al., 2006) grown on Ge (111) substrate using molecular beam epitaxy (MBE) were analyzed for structural and magnetic properties. Films with the dopant (Mn) concentration varying from 2% to 10% were fabricated (Ayoub et al., 2006). Presence of GeMn alloy clusters have been observed for higher dopant (Mn) concentrations. The formation of Ge_3Mn_5 precipitates on the layer was found to influence the ferromagnetic property of the thin film. Ferromagnetic behavior was found to increase as the dopant (Mn) concentration was increased in the sample of $Ge_{0.98}Mn_{0.02}$ (Ayoub et al., 2006).

Magnetic and electronic properties of $Ge_{1-x}Mn_x$ thin films grown using molecular beam epitaxy on SiO_2/Si (100) substrate exhibited p-type carriers and room temperature ferromagnetism (Ihm et al., 2004). It was also inferred from the SQUID measurements that, Ge_3Mn_5 phase present in the film was responsible for the room temperature ferromagnetism (Ihm et al., 2004) of the grown $Ge_{1-x}Mn_x$ thin films.

Crystalline $Si_{0.95}Mn_{0.05}$ thin films were prepared by post-thermal treatment of as deposited amorphous $Si_{0.95}Mn_{0.05}$ films and they were characterized for their magnetic and electrical properties (Zhang et al., 2004). Temperature dependence of magnetization measured using SQUID revealed that, the fabricated films exhibited ferromagnetic behavior at a temperature of $T_c > 400$ K. The X-ray diffraction analysis of them (Zhang et al., 2004) revealed that, the fabricated Si:Mn films were fully crystallized and the dopant (Mn) was incorporated in the host crystalline silicon lattice. Thermally activated conduction process was observed in the films through measurements of electrical properties (Zhang et al., 2004).

In polycrystalline $Si_{1-x}Mn_x$ (x = 0.005, 0.01, 0.015) prepared by Ma et al., (2006) using the arc melting method, it was observed that, the crystalline structure of silicon remains unaffected by the doping process when the concentration of Mn was low. Magnetic measurements revealed that the observed ferromagnetism has the transition temperature proportional to the concentration of the dopant. Metal-insulator transition was observed near the transition temperature for all the doped samples (Ma et al., 2006).

Bolduc et al., (2005) adopted the ion implantation method for preparing Mn implanted Si material and characterized the prepared material for its magnetic properties. The saturation magnetization of the annealed sample was found to increase. Significant differences in the temperature dependent remnant magnetization between the implanted p-type and n-type silicon were observed. Room temperature ferromagnetism was observed in all the prepared samples of Si:Mn.

Liu et al. (2006) prepared magnetron co-sputtered Mn doped Si films and analyzed the films for their structural, morphological and magnetic properties. It was found that, the as prepared film of Si:Mn was amorphous, while annealed film (800°C) was crystalline in nature. No isolation of Mn atoms was observed in the annealed film, while a granular feature uniformly covering the film surface was observed. Room temperature ferromagnetism was observed in the annealed films of Si:Mn (Liu et al., 2006).

Thin films of $Si_{1-x}Ni_x$ and $Si_{1-x}Cr_x$ prepared by electron beam evaporation method by Möbius et al., (1999) were found to exhibit metal-insulator transition. The prepared thin film samples were found to be amorphous and the conductivity of the DMS system $Si_{1-x}Ni_x$ was found to be affected drastically by the secondary phase of Si_2Ni present in the system.

The sample of $a\text{-}Si_{1-y}Ni_y{:}H$ prepared by RF sputtering in argon/hydrogen atmosphere by Bayliss et al., (1991) was found to be highly amorphous. From the electrical conductivity measurements recorded by them (Bayliss et al., 1991), it was found that semiconductor to metal (S-M) transition is possible with the increase in the dopant concentration (up to

26%). The DC conductivity measured at room temperature was found to increase from 10^{-5} $ohm^{-1}cm^{-1}$ to 10 $ohm^{-1}cm^{-1}$ as the dopant concentration was increased to the maximum limit. This increase in the DC conductivity of Si:Ni also evidenced the S-M transition in the system.

In another significant work by Collver (Collver, 1977), $Si_{1-x}Ni_x$ thin films prepared by electron beam evaporation revealed that, metastable impurity band has been found to be introduced due to the injection of Ni impurity into the host which resulted in S-M transition. The results of these works kindle interest to go further and analyze the properties of the materials in an effective way.

1.5   Methods of preparation of bulk DMS materials

There are many methods by which diluted magnetic semiconductors (DMS) can be prepared. Out of them, some commonly adopted methods are as follows;

  I. Mechanical alloying

 II. High temperature melt techniques

1.5.1 Mechanical alloying

Mechanical alloying is a process in which, the elemental powders of the semiconductor host and the dopant are mixed vigorously, in order to obtain the desired DMS compound. Mechanical alloying is done using a ball mill, which is a grinding instrument having a container in which, the elemental powders are loaded along with the grinding medium. The grinding medium may be balls made up of stainless steel, ceramic, agate, silicon carbide, flint pebbles, *etc*. High energy ball mills can grind the samples down to the nanometer scale. Mechanical alloying, used for grinding the particles to a smaller size, can also be used for the preparation of materials with enhanced physical and mechanical properties. Alloys and powders that are difficult to obtain in conventional melting and casting techniques can be produced using this method. Mechanical alloying is a unique process of preparation of alloys at room temperature. During the process of mechanical alloying, several factors are to be considered *viz.*, type of mill, type and materials of milling media, milling media to powder ratio, milling time, *etc*.

There are many types of ball mills such as high energy ball mill, planetary ball mill, tumbler rod mill, vibratory ball mill, laboratory ball mill, *etc*. The planetary ball mill and the tumbler rod mill are mainly used for industrial purposes. The high energy ball mill and the laboratory ball mill are used for grinding materials to a smaller size.

1.5.2 High temperature melt technique

The melt technique is another important technique used to prepare diluted magnetic semiconducting (DMS) materials. In this method, powders of the elements required to prepare the DMS material must be weighed according to the stoichiometric ratio and loaded in suitable crucibles, which can withstand the desired melting temperatures of the samples. A vacuum pumping system has to be used to evacuate the air in the crucibles containing the samples to avoid any possible oxidization of the samples. The evacuated crucible tubes must be vacuum sealed to avoid air entering in the crucibles. Then, the crucibles containing the samples are kept in a microcontroller controlled furnace and the temperature has to be set for the processing of the materials. Normally, the temperature is chosen just above the melting point of the chosen host material. This set temperature has to be increased in steps at regular intervals of time. When the set temperature is reached, the sample has to be kept at that temperature for soaking, to ensure the interaction of the host and the dopant atoms. After the soaking time, the melt has to be cooled slowly in small steps of temperature much slower than the heating cycle. This will reduce any possible strain during the cooling cycle. When the temperature of the furnace reaches room temperature, the processed samples are taken out and characterized for various properties.

1.6 Characterization techniques

1.6.1 Structural properties of materials using X-ray diffraction technique

It is essential to analyze the structural properties of a material since all the properties of a material depend on the atomic arrangement inside the material. Information on periodical arrangement of atoms in the lattice, bond angles and lengths, type of bonding, addition of impurities and their impact on the structural parameters, *etc.*, can be extracted from the observed X-ray diffraction pattern of the material.

Atoms in a crystal are arranged in a periodic manner symmetrically in three dimensional space. The lattice planes in the crystal act as the rulings in the grating and the distance between these planes are of the order of 10^{-10} m which is the same as that of the wavelength of the X-rays. Hence, when irradiated, X-rays are diffracted by the atoms in the lattice planes and the diffracted X-ray intensities are collected and analyzed. The diffraction of the X-rays by the lattice planes is illustrated in figure 1.5. The intensity of the observed X-rays will be maximum when the interference between two reflected rays is constructive and the Bragg equation is satisfied.

$$2d \sin \theta = n\lambda \tag{1.1}$$

where n is an integer, λ is the wavelength of the incident X-ray, θ is the scattering angle and d is the interplanar spacing.

The schematic diagram of X-ray diffractometer is given in figure 1.6. The essential components of X-ray powder diffractometer are (i) X-ray source which is normally a sealed X-ray tube (ii) goniometer which allows exact movements of the sample holder (iii) detector to receive the diffracted X-rays. In the X-ray tube, the X-rays are generated by the target (usually copper) and passed through collimating slits and directed to the powder sample placed in the goniometer. The rays which satisfy the Bragg condition are diffracted by the powder sample and will be received at the counter end after passing through the receiving slits. The purpose of the slits is to collimate the diffracted beam so that maximum information is obtained with minimum loss. The data is collected for various angles of diffraction in equal intervals. Thus, the obtained data is plotted for intensity counts against the diffraction angles and is called X-ray diffractogram.

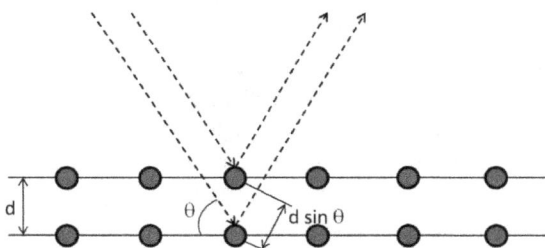

Figure 1.5 Diffraction of X-rays from lattice planes.

1.6.2 Morphological analysis

Morphological analysis in materials science is the study of shape, size, texture and phase distribution of physical objects. Materials having unique shapes and sizes may find various novel applications. When the size of materials is very small, *i.e*, in micro or nano scales, visible light radiation cannot be used for studying the surface of the materials. Hence, electron microscopes such as scanning electron microscope (SEM), transmission electron microscope (TEM) are used since electrons have short wavelengths which enable us to have observations with atomic resolution.

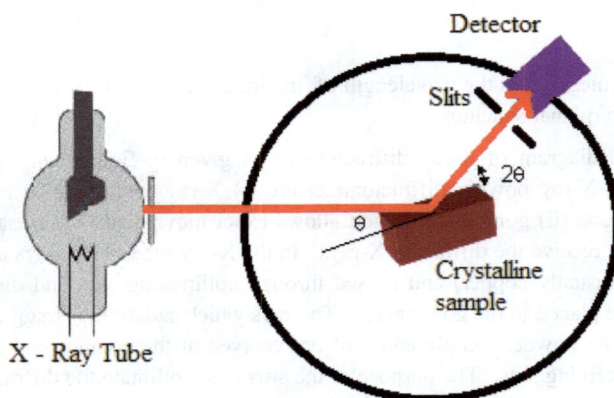

Figure 1.6 Schematic diagram of X-ray diffractometer.

1.6.2.1   Scanning electron microscopy

Scanning electron microscopy (SEM) is used to image the sample surface by scanning the sample with an electron beam in a rectangular pattern of parallel scanning lines. SEM uses a narrow electron beam resulting in high depth of field thus giving a three dimensional appearance of the surface of the sample under study.

The schematic diagram of a scanning electron microscope is given in figure 1.7. The main components of a scanning electron microscope are electron source, lenses, scanning coil, sample chamber and SEM detectors. Electrons are produced at the source by thermionic heating which are then accelerated and condensed into a narrow beam. This narrow electron beam is used for obtaining the SEM micrographs of materials. Tungsten filament is commonly used as the electron source. The spot on the contact surface will be smaller if the beam is narrower. After the beam is focused, the scanning coils deflect the beam in the X and Y axes so as to scan in the raster pattern. The SEM detectors collect the electrons coming from the samples.

Dried samples are used for examination using SEM in order to preserve the structure of the surface and to prevent cells from collapsing when exposed to high vacuum. The dried samples are mounted and coated with a thin layer of metal to avoid the charge generation on the surface. Thus, a clear image of the surface of the material is obtained. When the narrow beam of electron hits a particular area, the surface atoms discharge secondary electrons which are trapped by the detector. These secondary electrons strike a scintillator which emits light flashes that enter a photo multiplier. The light signal is then converted to an electrical current and amplified. This amplified signal is sent to a cathode ray tube where the image is produced which can be photographed. During the scanning process, most of the secondary electrons enter the detector directly while few reach the detector as back scattered electrons. Hence, raised areas appear lighter and depressions are darker which yield a three dimensional image of the surface of the sample. The SEM micrograph of a kidney stone is given as an example in figure 1.8.

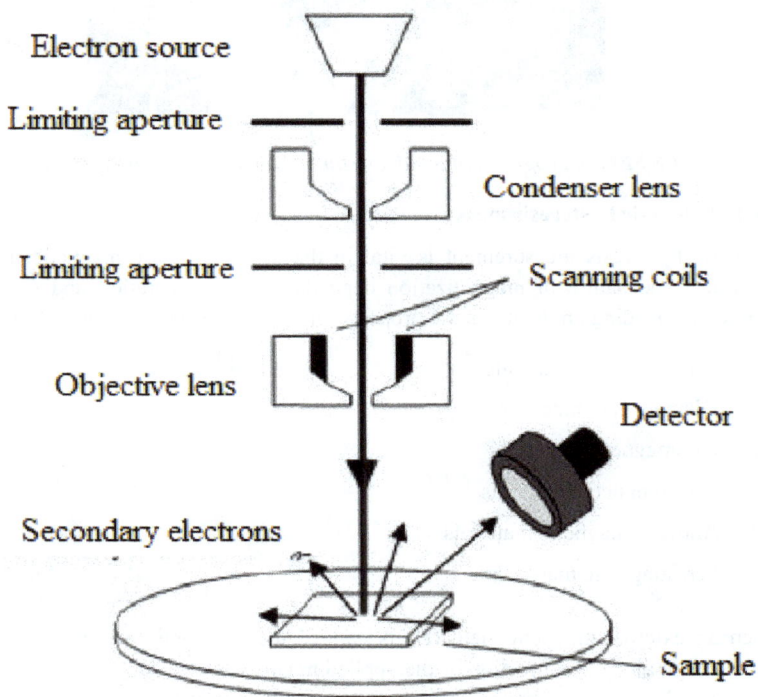

Figure 1.7 Schematic diagram of a scanning electron microscope.

Dilute Magnetic Semiconducting (DMS) Materials, R. Saravanan   Materials Research Forum LLC
Materials Research Foundations **35** (2018)   doi: http://dx.doi.org/10.21741/9781945291777

Figure 1.8 SEM micrograph of a kidney stone showing the 3D image of the surface.

1.6.3 Magnetic hysteresis measurements

Magnetic hysteresis measurement is vital in the analysis of the magnetic nature of a material. The nature of magnetization depends on the preparation and nature of the sample. Depending on the magnetic property, materials are classified as follows:

i.   Non magnetic materials

ii.   Diamagnetic materials

iii.   Paramagnetic materials

iv.   Ferromagnetic materials

v.   Antiferromagnetic materials

vi.   Ferrimagnetic materials

When an external magnetic field (H) is applied to a magnetic material, the magnetic dipoles align along the direction of the applied magnetic field. Due to the application of magnetic field, the material acquires magnetic property either temporarily or permanently. This process is called magnetization (M). If the alignment of the magnetic dipoles in the material is retained even after the removal of the external magnetic field, then, the material is said to be magnetized. To magnetize a material, applied magnetic field (H) is increased in the positive direction and to demagnetize the material, applied magnetic field (H) is reversed. The values of the magnetization (M) are plotted against

Dilute Magnetic Semiconducting (DMS) Materials, R. Saravanan   Materials Research Forum LLC
Materials Research Foundations **35** (2018)   doi: http://dx.doi.org/10.21741/9781945291777

the applied field strength (H) and this plot is called the hysteresis curve. A typical hysteresis curve is shown in figure 1.9.

When the field strength is increased in the positive direction, a state is reached above which the value of magnetization (M) does not increase further which is called magnetic saturation. During the demagnetization cycle, when the applied magnetic field is made zero, there will be a residual magnetization left in the material after the external magnetic field is removed. The measure of residual magnetization is called remanence. To completely demagnetize the material, the field H is increased in the negative direction. The measure of the ability of a material to withstand the applied magnetic field without being demagnetized is called the coercivity. All these parameters of the hysteresis curve determine the magnetic nature of the material under study. These measurements can be accomplished using a vibrating sample magnetometer.

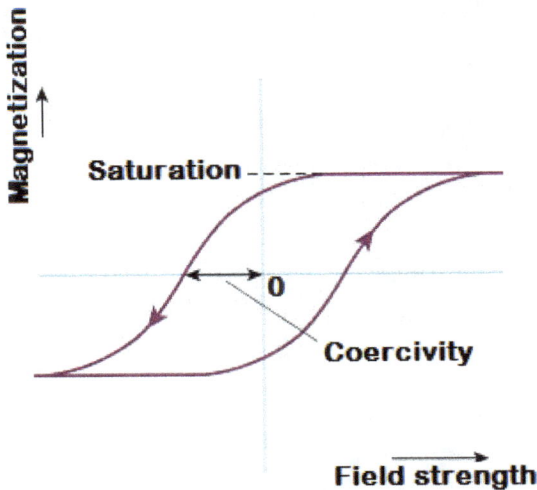

Figure 1.9 Hysteresis curve.

1.6.3.1   Vibrating sample magnetometer

Vibrating sample magnetometer (VSM) is an instrument used to measure magnetic properties of a given material as a function of magnetic field, temperature and time. Magnetic properties of powders, solids, liquids, single crystals and thin films can be analyzed using vibrating sample magnetometer. The VSM has high sensitivity and accuracy.   In a VSM, the sample is placed between pick up coils and vibrated mechanically to a sinusoidal motion. The change in magnetic flux caused due to this sinusoidal motion induces a voltage in the sensing coils.   The induced voltage is proportional to the magnetic moment of the sample. The magnetic field is applied by means of an electromagnet. The measurements can be recorded at various temperatures using cryostats or furnace assemblies.

The schematic diagram of a vibrating sample magnetometer is given in figure 1.10. The vibrating sample magnetometer consists of a sample holder in which the sample is placed centered in a pair of pick up coils between the poles of an electromagnet. The magnetic material placed on the sample holder is attached with the vibrator. The end of the rod is positioned between the pole pieces of an electromagnet. The vibrator moves the sample up and down and sets the sample frequency at around 85 Hz. The sample produces an alternating current at the same frequency as the vibration of the sample. The signal in the sensor coil is amplified using an amplifier and the signal generated contains the information about the magnetization of the sample. The resultant data can be graphed and plotted using software.

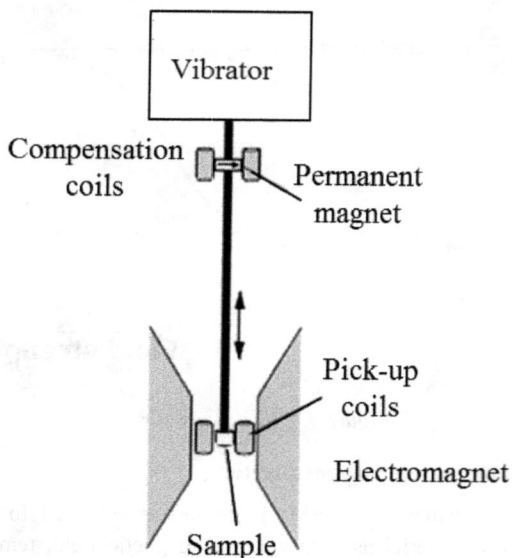

Figure 1.10 Schematic of a vibrating sample magnetometer.

References

[1]   Anthony R. West, Solid State Chemistry and its Applications, John Wiley & Sons, 1987.

[2]   Ayoub J.P., Favre L., Ronda A., Berbezier I., De Padova P., Olivieri B., Materials Science in Semiconductor processing, 9, 832 (2006). https://doi.org/10.1016/j.mssp.2006.08.055

[3]   Bayliss S.C., Asal R., Davis E.A., West T., J. Phys. Condens. Matter, 3, 793 (1991). https://doi.org/10.1088/0953-8984/3/7/004

[4]   Bolduc M., Affouda C.A., Stollenwerk A., Huang M.B., Ramos F.G., Agnello G., LaBella V.P., Phys. Rev. B, 71, 033302 (2005). https://doi.org/10.1103/PhysRevB.71.033302

[5]   Burch K., Awschalom D., Basov D., J. Magn. Magn. Mater. 320, 3207 (2008). https://doi.org/10.1016/j.jmmm.2008.08.060

[6]   Cho S., Choi S., Hong S.C., Kim Y., Ketterson J.B., Kin B.J., et al. Physical Review B 66, 033303 (2002). https://doi.org/10.1103/PhysRevB.66.033303

[7]   Collins D.M. Nature, 298, 49 (1982). https://doi.org/10.1038/298049a0

[8]   Collver M.M., Solid State Communications, 23, 333 (1977). https://doi.org/10.1016/0038-1098(77)91340-0

[9]   Datta S., Das B., Applied Physics Letters, 56, 665 (1990). https://doi.org/10.1063/1.102730

[10]   Farrow C.L., Juhas P., Liu J.W., Bryndin D., Bozin E.S., Bloch J., Proffen T., Billinge S.J.L., J. Phys.:: Condens. Matter. Phys. 19, 335219 (2007). https://doi.org/10.1088/0953-8984/19/33/335219

[11]   Furdyna J.K., J. Appl. Phys., 64, R29 (1988). https://doi.org/10.1063/1.341700

[12]   Gyeong S. Hwang, Margaret Debrowolska, Furdyna J.K., Taeghwan Hyeon, Nature Materials, 9, 47 (2010). https://doi.org/10.1038/nmat2572

[13]   Haynes W.M., CRC Handbook of Chemistry and Physics, CRC Press/Taylor and Francis, Boca Raton, FL, 95th Edition, 2014.

[14]   Ihm Y.E., Yu S.S., Jim D., Kim H., Lee K.H., Kim C.S., et al. Journal of Magnetism and Magnetic Materials e1539, 272 (2004).

[15]   Izumi F., Dilanian R.A., Recent Research Developments in Physics, Vol. 3, Part II, Transworld Research Network, Trivandrum, 2002.

[16] Jeong I.K., Thompson J., Proffen T., Perez A., Billinge S.J.L., J. Appl. Cryst., 34, 536 (2001). https://doi.org/10.1107/S0021889801009207

[17] Johnson M., Bennett B., Yang M., IEEE Trans. Magn., 34, 1054 (1998). https://doi.org/10.1109/20.706355

[18] Kittel C., Introduction to Solid State Physics, 5th Ed, John Wiley & Sons Inc., 1976.

[19] Lee S., Dobrowolska M., Furdyna J.K., Luo H., Ram Mohan L.R., Phys. Rev. B., 54, 16939 (1996). https://doi.org/10.1103/PhysRevB.54.16939

[20] Li A.P., Shen J., Thompson .J.R., Weitering H.H., Applied Physics Letters 86, 152507 (2005). https://doi.org/10.1063/1.1899768

[21] Liu L., Chen N., Wang Y., Yin Z., Yang F., Chai C., Zhang X., Journal of crystal growth, 291, 239 (2006). https://doi.org/10.1016/j.jcrysgro.2006.02.033

[22] Ma S.B., Sun Y.P., Zhao B.C., Tong P., Zhu X.B., Song W.H., Solid state communications 140, 192 (2006). https://doi.org/10.1016/j.ssc.2006.07.039

[23] Möbius A., Frenzel C., Thielsch R., Rosenbaum R., Adkins C.J., Schreiber M., Bauer H.D., Grötzschel R., Hoffmann V., Krieg T., Matz N., Vinzelberg H., Witcomb M., Phys. Rev. B, 60, 14209 (1999). https://doi.org/10.1103/PhysRevB.60.14209

[24] Momma K., Izumi F., J. Appl. Crystallogr., 44, 1272 (2011). https://doi.org/10.1107/S0021889811038970

[25] Morresi L., Pinto N., Ficcadenti M., Murri R., D'Orazio F., Lucari F., Materials Science and Engineering B 126, 197 (2006). https://doi.org/10.1016/j.mseb.2005.09.025

[26] Munekata H., Ohno H., von Molnar S., Segmuller A., Chang L.L., Esaki L., Phys. Rev. Lett. 63, 1849 (1989). https://doi.org/10.1103/PhysRevLett.63.1849

[27] Okabayashi J., Kimura A., Rader O., Mizokawa T., Fujimori A., Hayashi T., Tanaka M., Phys. Rev. B., 58, R4211 (1998). https://doi.org/10.1103/PhysRevB.58.R4211

[28] Ohno H., Science, 281, 951 (1998). https://doi.org/10.1126/science.281.5379.951

[29] Ohno H., Shen A., Matsukura F., Oiwa A., Endo A., Katsumoto S., Iye Y., Appl. Phys. Lett. 69, 363 (1996). https://doi.org/10.1063/1.118061

[30] Ohring M., "Reliability and failure of electronic materials and devices", Academic Press, 1998, ISBN 0-12-524985-3.

Dilute Magnetic Semiconducting (DMS) Materials, R. Saravanan   Materials Research Forum LLC
Materials Research Foundations 35 (2018)   doi: http://dx.doi.org/10.21741/9781945291777

[31] Park Y.D., Hanbicki A.T., Erwin S.C., Hellberg C.S., Sullivan J.M., Mattson J.E.,
      et al. Science 295, 651 (2002). https://doi.org/10.1126/science.1066348

[32] Peaton S.J., Abernathy C.R., Norton D.P., Hebard A.F., Park Y.D., Boatner L.A.,
      Budai J.D., Materials Science and Engineering R, 40, 137 (2003).
      https://doi.org/10.1016/S0927-796X(02)00136-5

[33] Petříček V., Dušek M., Palatinus L., Zeitschrift für Kristallographie-Crystalline
      Materials, 229 (5), 345 (2014). https://doi.org/10.1515/zkri-2014-1737

[34] Pohm A., Comstock C., Hurst A., J. Appl. Phys., 67, 4881 (1990).
      https://doi.org/10.1107/S0021889899003532

[35] Proffen T., Billinge S.J.L., Journal of Applied Crystallography, 32, 572 (1999).
      https://doi.org/10.1107/S0021889899003532

[36] Rietveld H.M., Journal of Applied Crystallography, 2, 65 (1969).
      https://doi.org/10.1107/S0021889869006558

[37] Savchuk A.I., Stolyarchuk I.D., Makoviy V.V., Savchuk O.A., Applied Optics, 53
      (10), 822 (2011).

[38] Tsymbal E.Y., Pettifor D., Solid State Physics 56, 113 (2001).
      https://doi.org/10.1016/S0081-1947(01)80019-9

[39] Vittorio M.N.P., Corrado de Tullio, Benedetto T., Mario L.N., Giovanni G. and
      Fransesco D.L., Sensors, 12, 15558 (2012). https://doi.org/10.3390/s121115558

[40] Wilamowski Z., Werpachowska A.M., Mater. Sci. Pol., 24, 3 (2006).

[41] Zhang F.M., Applied Physics Letters 85, 786 (2004).
      https://doi.org/10.1063/1.1775886

Dilute Magnetic Semiconducting (DMS) Materials, R. Saravanan   Materials Research Forum LLC
Materials Research Foundations **35** (2018)   doi: http://dx.doi.org/10.21741/9781945291777

Chapter 2

Analytical Techniques

Abstract

Chapter 2 introduces the analytical techniques used to analyze the charge density of the proposed DMS materials. Analysis based on the obtained powder X-ray diffraction intensities is explained in this chapter. The detailed theory of charge density analysis has been presented along with the mathematical tools used for them. An introduction on the analysis of the structural properties and electronic charge density distribution of materials is given in this chapter. Also, analysis of the local structure of materials is elaborated. A review on some earlier works done on the charge density analysis of materials is also elaborated here.

Keywords

Powder Data, Rietveld, Charge Density, Pair Distribution Function, Maximum Entropy Method, Fourier Method, Local Structure, Structure Factor

Contents

2. Analytical techniques

Analyzing the characteristics of materials is vital for understanding the properties of materials and to understand the structure of the material, the spatial arrangement of atoms in the material, the distribution of electronic charges, etc. The properties of a chosen material can be analyzed using various sophisticated analytical techniques. One of the most important techniques is the X-ray diffraction technique to solve the structure and to extract structure related information about the materials. Some of the techniques which employ the observed X-ray diffraction data of the samples are as follows;

a) Rietveld refinement technique

b) Charge density estimation by

    (i) Fourier method (Coppens, 1997)

    (ii) Maximum entropy method (Collins, 1982)

c) Local structure analysis by pair distribution function (Proffen et al., 1999)

2.1 Rietveld refinement technique

Rietveld refinement technique (Rietveld, 1969) is very useful to characterize the powder materials for their structural properties. This refinement technique was devised by Hugo Rietveld (Rietveld, 1969). To obtain the diffraction pattern of crystalline samples, radiations like X-rays and neutron are used. The observed diffraction pattern contains intensity peaks whose height, width and position are used to determine the structure of the material under study.

The Rietveld method (Rietveld, 1969) uses the least squares approach to refine a theoretical line profile until it matches the experimentally measured profile. This method can be used to deal with strongly overlapping reflections. This method was initially used for the diffraction of monochromatic neutrons where positions of the reflections are given in terms of 2θ, the Bragg angle. The diffraction technique can be implemented to both single crystals and powder samples. In most novel materials, it is very difficult to grow large single crystals and hence powders of the samples are preferred for structure analysis where the material is in the form of very small crystallites. The peaks of the conventional powder diffraction pattern overlap due to various reasons like symmetry conditions, accidental overlapping due to limited experimental resolution and background radiation. Hence, it is difficult to define with accuracy, non-random distribution of crystallites in the specimen. Hence, correction for preferred orientation is used when there is a stronger tendency for the crystallites to be oriented more in a particular way. The overlapping of the diffraction peaks makes the structure determination difficult. But, the Rietveld method (Rietveld, 1969) creates a virtual separation of the overlapping peaks, thus making the structure determination accurate.

Rietveld refinement (Rietveld, 1969) method is a technique in which user-selected parameters are refined to minimize the difference between an experimental X-ray diffraction pattern and a model based on the hypothesized crystal structure and instrumental parameters. This method is used to fit the full profile using crystallographic constraints. The peak positions are constrained using lattice parameters and space group. The peak intensities are constrained using the crystal structure.

Rietveld refinement method (Rietveld, 1969) can be used to fit not only the integrated intensities but the whole diffraction profile. The quantitative analysis for both crystalline and amorphous materials can be done using this method. Refinement of structural parameters such as lattice parameters, atomic positions, occupancies and temperature vibrations both isotropic and anisotropic can be done in this method. Information on grain size, isotropic and anisotropic micro strain, stacking and twin faults and magnetic moments can also be obtained using the Rietveld refinement method (Rietveld, 1969). However, this method is not intended for solving the structure of a chosen material. In fact, the structure model of the material under study must be known before starting the refinement procedure.

There are some basic requirements in order to carry out the refinement of the observed X-ray diffraction profile of the material under study. The requirements are as follows:

a) A high quality experimental diffraction pattern is needed.

b) A structure model is required that makes physical and chemical sense.

c) Suitable peaks and background functions are required.

In the refinement process, the experimental intensities are compared with those of the calculated ones till a minimum difference is obtained. An X-ray diffractogram can be recorded for a sample, as a profile plotted in the form of intensity versus Bragg angle, 2θ. An example is given in figure 2.1.

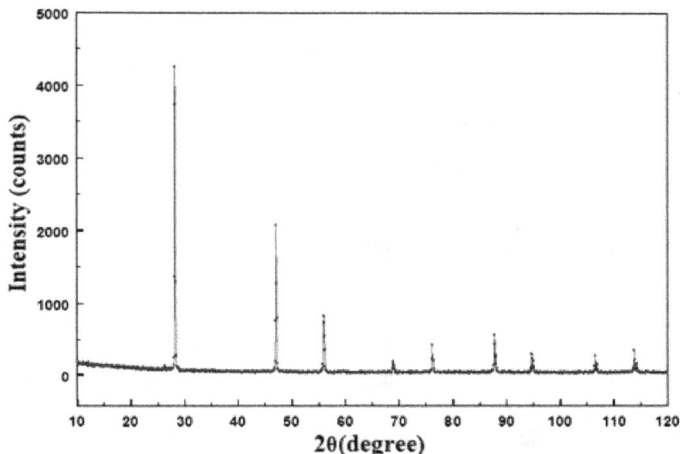

Figure 2.1 X-ray diffractogram of Silicon.

The X-ray diffraction pattern contains intensity distribution with respect to the Bragg angle. The calculated intensity at a given point I_k of the diffractogram is given by

$$y_{ic} = y_{ib} + \sum_{\Phi} S_{\Phi} \sum_k G_{\Phi} \left(2\theta_i - 2\theta_k\right) I_k, \tag{2.1}$$

where G is the normalized profile shape function, I is the intensity of the kth reflection, y_{ib} is the background and S is the scale factor of phase Φ. The summation is performed over all the phases Φ and all the reflections k contributing to the respective point.
The intensity of the Bragg reflection is given by

$$I_k = m_k L_k |F_k|^2 P_k A_k \tag{2.2}$$

Where m_k is the multiplicity of k, L_k is the Lorentz polarization factor, $|F_k|^2$ is the structure factor, P_k is the preferred orientation factor and A_k is the absorption factor
The ultimate goal of the refinement process is to minimize the residual function

Materials Research Forum LLC
doi: http://dx.doi.org/10.21741/9781945291777

$$\sum_i w_i \left(y_i^{obs} - y_i^{cal}\right)^2 \tag{2.3}$$

Where $w_i = 1/y_i^{obs}$, where y_i^{obs} is the observed intensity at the ith step and y_i^{cal} is the calculated intensity at the ith step.

The calculated intensity after i number of steps is

$$I_i^{cal} = S_F \sum_{j=1}^{N\ phases} \frac{f_j}{v_j^2} \sum_{k=1}^{M\ peaks} L_k \left|F_{k,j}\right|^2 S_j\left(2\theta_i - 2\theta_{k,j}\right)P_{k,j}A_j + bkg_i \tag{2.4}$$

Where $S_F \sum_{j=1}^{N\ phases} \frac{f_j}{v_j^2}$ is the scale factor in which S_F is the beam intensity, f_j is the volume fraction, V_j is the cell volume, L_k is the Lorentz polarization factor which consists of the geometry, orientation of the monochromator (angle θ), detector, beam size/ sample volume and sample positioning (angular) and $\left|F_{k,j}\right|^2$ is the structure factor consisting of the multiplicity of kth reflection (m_k) and the temperature factor.

Also, $P_{k,j}$ is the preferred orientation, A_j is the volume absorption, i is the number of steps, j is the number of phases and k indicates the kth reflection. Determination of structure factor $F_{k,j}$ is the most important thing as it contains the information about the system involved.

2.1.1 Structure factor

The wavelength of X-rays are comparable with atomic dimensions and hence there will be phase differences between the waves scattered by the electrons in different regions. If we consider a group of atoms in a unit cell of a crystal, $A_1, A_2,......., A_r$, whose positions defined with respect to the origin are the vectors $r_1, r_2, r_3,......., r_r$, the atoms have the scattering factors $f_1, f_2, ..., f_j$ respectively. If the group of atoms is considered as a unit, then, the equation for total amplitude is given by

$$A_j = A_0 \sum f_j exp\left(i\varphi_j\right) \tag{2.5}$$

Where φ_j is the phase shift of the jth atom due to scattering

A_0 is the amplitude at the source before scattering

A_j is the scattered amplitude due to the jth atom

f_j is the scattering factor of the jth atom

Dilute Magnetic Semiconducting (DMS) Materials, R. Saravanan   Materials Research Forum LLC
Materials Research Foundations **35** (2018)   doi: http://dx.doi.org/10.21741/9781945291777

The structure factor is then defined as

$$F_{hkl} = \sum_{j=1}^{m} N_j f_j \, exp\left[2\pi i \left(hx_j + ky_j + lz_j\right)\right] \tag{2.6}$$

And $|f|^2 = \left(f_0 \exp\left[\frac{-Bsin^2\theta}{\lambda^2}\right] + \Delta f'\right)^2 + (\Delta f'')^2$ $\tag{2.7}$

where B is the Debye-Waller factor.

In order to solve a crystal structure using Bragg's Law and F_{hkl}, the diffraction pattern from an ideal crystal can be calculated. Hence, it is possible to simulate ideal powder diffraction patterns if the parameters such as space group symmetry, unit cell dimensions, atom types, relative coordinates of atoms in the unit cell, atomic site occupancies, atomic thermal displacement parameters are known for each and every phase in the sample.

2.1.2 Procedure followed in Rietveld refinement process

The Rietveld refinement (Rietveld, 1969) method involves three steps viz., experiment, analysis and refinement. First and foremost, a correct instrument and appropriate experiment conditions should be selected. The sample should be prepared accordingly to collect the X-ray diffraction pattern. Data analysis is to be done prior to the refinement routine in order to verify the quality of the data and perform qualitative analysis.

The procedure for the refinement of the experimental diffraction pattern is as follows: The parameters such as cell dimension, intensities and background are adjusted manually. The background and overall intensities are refined followed by the positions and shapes of the peaks. The obtained crystal structures are refined and the results are assessed after a minimum difference is obtained between the observed and calculated profiles of the samples. Quantitative assessment is to be done by reliability indices. When the R-factors are as low as and < 4%, the refinement process can be considered successful.

There are some advantages of using the Rietveld refinement method (Rietveld. 1969) for refining the X-ray diffraction profiles of materials. The Rietveld refinement technique (Rietveld, 1969) uses the measured intensities of the materials. This method is less sensible to model errors and experimental errors. However, there are some shortcomings for this method. A perfect theoretical model is required for refinement to begin. Also, wide range of X-ray data is needed for obtaining the best results, with least possible errors.

2.2 Electronic charge density distribution

The ultimate aim of the X-ray experiment is to map charge density distribution of the molecule inside a unit cell. Hence, best possible methods to determine the charge density should be chosen. The importance of understanding charge density is outlined as follows;

The quantum mechanical probability density or simply the number density of electrons is called electron density. Atoms are present in molecules and the regions of electron density are found around the atoms and their bonds as electron clouds. Electron densities are usually given as isodensity surface or simply isosurface where the size and shape of the surface is determined by the percentage of total electrons enclosed in the particular volume under study. The graphical representation of the electronic charge density distribution gives an understanding of other electronic properties which depend on the distribution of electrons. The observation of the distribution of electrons enables us to actually visualize the chemical bonding between atoms. This is very helpful to analyze molecules and crystals in terms of electron density partitioning (Coppens, 1997). The electron density analysis can be applied to study the coordinative bonds (Macchi et al., 2003) and intermolecular interactions (Flaig et al., 2002).

The lattice of a crystal system has a periodicity and hence the electron density of the system will also be a periodic function. In a volume element dV, the number of electrons is $\rho(x,y,z)dV$. When the X-rays are used to scan the crystalline sample, the wavelet scattered by the volume element will be,

$$\rho(x, y, z)exp[-2\pi i(hx + ky + lz)]dV \tag{2.8}$$

If we consider the whole unit cell, the resultant sum of contributions from all such elements in the unit cell will be

$$F_{hkl} = \int \rho(x, y, z)exp\left[-2\pi i(hx + ky + lz)\right]dV \tag{2.9}$$

Addition of the waves scattered in the direction of the (*hkl*) reflection from the atoms in the unit cell results in the structure factor. This is achieved by assuming that the scattering power of the electron cloud surrounding each atom can be equated to the scattering power of the actual number of electrons concentrated at the atomic centre.

From the observed X-ray intensities in the diffractogram, the structural properties of the unit cell of the crystalline samples can be deduced. The relative intensities of the reflections are measured in order to deduce the electron density distribution in the unit cell of the crystal. The electron density distribution is connected to the intensities. The electron density distribution of the unit cell can be calculated if the phases and the structure factors are known. If a set of the structure factor is given, it essentially corresponds to one and only one electron distribution.

The square root of the measured diffraction intensity gives the magnitude of the individual structure factors and the phases are determined by solving the structure. This elucidation is taken as a model and improved by least squares refinement based on the structure factors to obtain an accurate model. By taking the Fourier summation of the phased structure factors, the electron density can be calculated. The constructive interference of the scattered X-rays from different atoms in the crystal structure results in the observed diffraction intensities in the X-ray diffractogram. The diffraction intensities are the Fourier transform of the crystal structure and thus the crystal structure is the Fourier transform of the diffraction intensities which is expressed in terms of electron density distribution concentrated in atoms. This cannot be measured by direct experimental methods, since the scattered X-rays cannot be refracted by lenses to form an image as done with light in an optical microscope. Also, it cannot be obtained directly by calculation since the relative phases of the waves are unknown. Hence, the electron density distribution can be calculated using the Fourier series given the set of structure factors.

2.2.1 Fourier method

Fourier series is a series of trigonometric terms using which one can represent any well-behaved continuous periodic function. In a small volume dV, the number of electrons is $\rho(r)dV$. For this small volume element dV, the scattering amplitude will be $\rho(r)dV$ times the scattering amplitude from an electron at the same position. Therefore, the total scattering amplitude can be calculated from the distribution of the electron density which is represented as $\rho(r)$. $F(H)$ is expressed in terms of density as

$$F(H) = \int \rho(r)\, exp(2\pi i H \cdot r) dV \tag{2.10}$$

The electron density is given by the inverse Fourier transform of the above equation as follows:

$$\rho(r) = \int F(H)\, exp(-2\pi i H \cdot r) dV = \frac{1}{V}\sum F(H)\, exp(-2\pi i H \cdot r) \tag{2.11}$$

The integral is replaced by the summation because $F(H)$ is defined at the discrete set of reciprocal lattice points k. If the structure factor is taken as $(H) = A(H) + iB(H)$, then the electron density can be written as

$$\rho(r) = \frac{1}{V}\sum[A(H) + iB(H)][\cos(2\pi H \cdot r) - i\sin(2\pi H \cdot r)] \tag{2.12}$$

Since the electron density is a real function, it is taken as $A(H) = A(-H)$ and $B(H) = B(-H)$. Thus, we have the electron density as

$$\rho(r) = \frac{1}{V}\sum_{1/2}[2A(H)\cos(2\pi H \cdot r) + 2B\sin(2\pi H \cdot r)] \tag{2.13}$$

Taking $A(H) = |F(H)|\cos\varphi$ and $B(H) = |F(H)|\sin\varphi$, the density becomes

$$\rho(r) = \frac{2}{V}\sum_{1/2}[|F(H)|\cos\varphi\cos(2\pi H \cdot r) + |F(H)|\sin\varphi\sin(2\pi H \cdot r)] \tag{2.14}$$

This can be further reduced as

$$\rho(r) = \frac{2}{V}\sum_{1/2}[|F(H)|\cos(2\pi H \cdot r) - \varphi(H)] \tag{2.15}$$

From the above equation, i.e., Eqn (2.15), we can understand that each structure factor contributes a plane wave to the total electron density with wave vector H and phase φ. Thus, to calculate the electron density, phases of the structure factors need to be known. If we know the approximation to the scattering density, then $\varphi(H)$ can be calculated and an imperfect image of the structure can be obtained. The period of the plane wave having the amplitude $F(H)$ in the direction of the wave vector H is $1/H$. Hence, for the higher order reflections, the period is shorter, thus adding resolution to the image. As more numbers of higher order reflections is added, we can get an improved image of higher resolution. As we cannot use lenses for X-ray beams, computational methods are essential to obtain the Fourier transform of the diffraction pattern as the image. To calculate the charge density using the Fourier method, an infinite number of Fourier coefficients are needed. But in practice, only a finite number of coefficients are available through experiment. Also, the experimental errors are neglected by setting the entire missing Fourier coefficient as zero which is a highly biased assumption. This gives rise to a negative electron density which is not physical and impedes the understanding of the

Dilute Magnetic Semiconducting (DMS) Materials, R. Saravanan · · · · · · · Materials Research Forum LLC
Materials Research Foundations 35 (2018) · · · · · · · doi: http://dx.doi.org/10.21741/9781945291777

intricate details in the valence region such as the bonding charge. A more practical way of establishing all positive charge density is by means of a versatile statistical tool that undermines any experimental noises. One such method is called the maximum entropy method (MEM), which is discussed below.

2.2.2 Maximum entropy method

To analyze the electronic properties of a particular material, an exact electron density distribution is needed. To fulfill this purpose, all the structure factors should be essentially known as found in Fourier and multipole formalisms. X-ray diffraction methods have their own limitations making the collection of the exact values of all the structure factors impossible. Practically, the number of the structure factors obtained by the experimental method is limited. Hence, merely taking the inverse Fourier transform with the limited number of data for the construction of the charge density will be inappropriate. For this situation, the maximum entropy method (MEM) serves as an appropriate technique to solve the problem. Here, in this method, the uncertainties are dealt with the concept of entropy.

This method is information-theory based technique used to enhance the information obtained from noisy data. This technique was originally developed to use in the field of radio astronomy (Gull et al., 1978). Later, this technique was found to be useful in solving the charge density distribution (Collins, 1982) and widely applied in the study of materials. The equations used in this method is built on the basis of statistical thermodynamics, where the particles are distributed over position and momentum space called phase space while in the information theory, the distribution of the numerical quantities are over the ensemble of pixels. The probability of a distribution of N identical particles over m boxes each populated by n_i particles is given by

$$P = \frac{N!}{n_1! n_2! n_3! \dots n_m!} \tag{2.16}$$

Using the Stirling's formula, the entropy is

$$S = -\sum_i n_i \ln n_i \tag{2.17}$$

For a prior probability q_i for box containing n_i particles, the probability is given by

$$P = \frac{N!}{n_1! n_2! n_3! \dots n_m!} q_1^{n_1} q_2^{n_2} \dots q_m^{n_m} \tag{2.18}$$

Hence, the entropy becomes,

$$S = -\sum_i n_i \ln n_i + \sum_i n_i \ln q_i = -\sum_{i=1}^{m} n_i \ln \frac{n_i}{q_i} \tag{2.19}$$

Using the entropy formula, the electron density distribution is expressed as a sum over M grid points in the unit cell as

$$S = -\sum \rho'(r) \ln \left(\frac{\rho'(r)}{\tau'(r)}\right) \tag{2.20}$$

$\rho'(r)$ and $\tau'(r)$ are the probability and prior probability respectively and are related to the actual electron density in the unit cell as

$$\rho'(r) = \frac{\rho(r)}{\sum_r \rho(r)} \quad \text{and} \quad \tau'(r) = \frac{\tau(r)}{\sum_r \tau(r)} \tag{2.21}$$

Where $\rho(r)$ and $\tau(r)$ re the electron and prior electron densities respectively at a certain fixed r in the unit cell. Instead of normalized densities actual densities are used and when there is no information $\rho'(r)$ equals $\tau'(r)$. The normalized densities are

$$\sum \rho'(r) = 1 \quad \text{and} \quad \sum \tau'(r) = 1 \tag{2.22}$$

A constraint is introduced into this as

$$C = \frac{1}{N} \sum_k \frac{|F_{Cal}(k) - F_{obs}(k)|^2}{\sigma^2(k)} \tag{2.23}$$

where N is the number of reflections used for the analysis, $\sigma(k)$ is the standard deviation of the observed structure factor $F_{obs}(k)$. The calculated structure factor $F_{Cal}(k)$ is given by

$$F_{Cal}(k) = V \sum \rho(r) \exp(-2\pi i \boldsymbol{k}.\boldsymbol{r}) dV \tag{2.24}$$

The constraint C is known as weak constraint. When the values of the calculated structure factors and the observed structure factors agree then the constraint C becomes unity. The Fourier transform of the electronic charge density distribution in a unit cell is the structure factor. Any kind of deformations of the electron densities may be permitted in real space if the information containing the deformations is included in the observed data.

While maximizing the entropy, the constraint C should become unity. For the calculations, Lagrange's method of undetermined multiplier λ is used. Then

$$Q = S - \left(\frac{\lambda}{2}\right) C = -\sum \rho'(r) \ln \left(\frac{\rho'(r)}{\tau'(r)}\right) - \frac{\lambda}{2N} \sum_k \frac{|F_{Cal}(k) - F_{obs}(k)|^2}{\sigma^2(k)} \tag{2.25}$$

When $\frac{dQ}{d\rho} = 0$ and $\ln x = x - 1$, then

$$\rho(r_i) = \tau(r_i) \exp \left\{\left(\frac{\lambda F_{000}}{N}\right) \left[\sum \frac{1}{\sigma(k)^2}\right] |F_{obs}(k) - F_{cal}(k)| \exp(-2\pi j k \cdot r)\right\} \tag{2.26}$$

where F_{000} is the total number of electrons in the unit cell which is the atomic number of the atom. In order to solve Eqn. (2.26) an approximation is introduced to replace $F_{cal}(k)$ as follows.

$$F_{cal}(k) = V \sum \tau(r) \exp(-2\pi i k \cdot r) \, dV \tag{2.27}$$

This approximation is called the zeroeth order single pixel approximation. Applying this to the right hand side of Eqn. (2.26) helps in solving the equation in an iterative way if the initial density for the prior distribution is given. A uniform density distribution is given as the prior density $\tau(r)$ as $0 \leq \tau(r) \geq Z/M$ where M is the total number of pixels for which the electron density is calculated. The uniform density distribution indicates a minimum entropy state and hence it is assigned for the prior density. In the charge density calculations, the total number of electrons in the unit cell is kept constant throughout till all the symmetry elements are satisfied. A limited number of pixels are used in the numerical calculations and hence summation is used in all the equations (Eqn. 2.17 to 2.27).

A high resolution density distribution can be obtained using the maximum entropy method (MEM) (Collins, 1982) from a limited number of diffraction data without using a structural model. Maximum entropy method (MEM) (Collins, 1982) is a statistical deduction method suitable for the examination of electron densities in the inner atomic region, like bonding region. The MEM method gives least biased information on the electron densities. Before starting the MEM procedure necessary corrections should be made for the extinctions present in the observed data set. Also, corrections should be made for scattering for the structure factors. A model is required for the evaluations of these corrections for which the independent atom model is sufficient.

The maximum entropy method has significant advantages over the other methods of charge density calculation as follows: Instead of normalized electron density, an explicit formulation for the actual electron density is given in the MEM method. Also, the performance of the calculation method is accurate, even when the available information is limited. MEM method is a least biased calculation method, which gives accurate information of the structure factor of a material. The electron density distribution of a material can be mapped very precisely using the MEM method and hence, the existence of bonding electrons can be clearly seen.

2.3   Local structure analysis

Another practical way of analyzing the presence of any possible defects in the structure is by employing local structure analysis. The study of crystalline materials highlights the role of the periodic arrangement of atoms and its fundamental symmetry, which has effects on the description of properties of the crystalline materials. But, when the studies are focused on highly amorphous materials like glasses and liquids, long range symmetry and periodicity are ruled out and the only thing one can focus is the arrangement of neighboring group of atoms, which is called the local structure. The local structure analysis plays a vital role in analyzing materials like negative thermal expansion materials, pharmaceutical materials, optoelectronic materials, electrically or magnetically ordered materials such as shape memory alloys. The important method of determining the local structure of a material is total scattering analysis, in which, X-rays are used for collecting the scattered intensity data. Generally, a Fourier transform is performed for the entire spectrum of the scattered intensity thereby providing a histogram of all interatomic distances. The local structure analysis can also be done for crystalline solids since the properties of the crystalline substances are mostly defined by the short-range fluctuations of the crystal structure. Hence, the analysis of the local structure finds an important role in materials science.

2.3.1   Pair distribution function

The local structure of the materials is accomplished using the pair distribution function (PDF) (Proffen et al., 1999) analysis. Pair distribution function is a method, that describes the probability of finding any two atoms at a given inter-atomic distance r. The application of the PDF is for the materials with short-range periodicity and hence, it shows broad features and does not extend over the short-range of the first few

Dilute Magnetic Semiconducting (DMS) Materials, R. Saravanan   Materials Research Forum LLC
Materials Research Foundations **35** (2018)   doi: http://dx.doi.org/10.21741/9781945291777

coordination spheres.   The pair distribution function is calculated from the Bragg intensities, as well as, the diffuse scattering intensities. Hence, it can distinguish between short range order and random displacements of the atoms.

The experimental pair distribution function $G(r)$ is directly obtained from the diffraction data, by Fourier transforming the normalized total structure factor $S(Q)$ with

$$Q = 4\pi(\sin\theta)/\lambda \qquad (2.28)$$

where $S(Q)$ is the measured intensity corrected for background, Compton scattering, multiple scattering, absorption, geometric and other factors.
The pair distribution function is given as

$$G(r) = 4\pi r[\rho(r) - \rho_0] = \left(\frac{2}{\pi}\right)\int_0^\infty Q[S(Q) - 1]\sin(Qr)\,dQ \qquad (2.29)$$

where $\rho(r)$ is the microscopic pair density, ρ_0 is the average number density and Q is the magnitude of the scattering vector.   The factors influencing the quality of the experimental pair distribution function are the Fourier termination errors, resolution and counting statistics.   The termination errors can be minimized by measuring to largest values of Q which is limited by the instrument setting for the highest diffraction angle and the wavelength of the X-ray source.

For a pair of atoms i and j within a crystal, for which the pair distribution is to be calculated can be done from a structural model as,

$$G_c(r) = (1/r)\sum_i\sum_j\left[\left(\frac{b_i b_j}{\langle b\rangle^2}\right)\delta(r - r_{ij})\right] - 4\pi r\rho_0 \qquad (2.30)$$

where r_{ij} is the separation between the atoms, b_i is the scattering power of atom i and $\langle b\rangle$ is the average scattering power of the sample. The thermal displacements are introduced by displacing atoms in the model to a given isotropic Debye-Waller factor.   So, as an alternative procedure, each delta function in Eqn.(2.30) can be convoluted with a Gaussian accounting for the displacements. The width σ'_{ij} of the Gaussian is given by the anisotropic thermal factors $U_{lm} = \langle u_l u_m\rangle$ of i and j atoms. The experimental width σ_{ij} is dependent on the separation distance between the atoms (Jeong et al., 1998) and the relation is given as

$$\sigma_{ij} = \sigma'_{ij}(p) - \left(\delta/r_{ij}^2\right) \qquad (2.31)$$

The interpretation of the experimental pair distribution function is similar to the Rietveld (Rietveld, 1969) method. As the origin to calculate the PDF, a structural model of the atomic arrangement is considered. By the optimization of the model parameters, a PDF which is in good agreement with the experimental PDF is obtained and then analyzed.

2.4 A review of literature on charge density analysis

Considerable work has been done on the analysis of the charge density of materials using Fourier techniques, multipole formalism and maximum entropy method. The studies have uncovered many interesting results regarding charge transfer, bonding and magnitude of the bond strength. These studies are very important in analyzing the properties of semiconductor based materials, which in turn is very helpful in fabricating the materials for various purposes.

Attard and Azaroff (1963) carried out an analysis on indium antimonide (InSb), which revealed the covalent nature of the bonding between indium and antimony atoms. In a work done by Dawson (1967), the mid-bond densities of diamond, silicon and germanium were observed as 0.64 e/Å^3, 0.25 e/Å^3 and 0.18 e/Å^3 respectively, using X-ray measurements and the results show that, the bonding between the atoms is covalent (Dawson, 1967 a,b,c).

The electronic structure of CaF_2 was determined using MEM and the mid-bond density was found to be 0.69 e/Å^3 at a distance of 1.33Å from Ca atom situated at the origin. The nature of the bonding was found to be ionic (Saravanan et al., 2004).

In another work done by Saravanan et al., (2005), the charge density distribution of ZnTe was studied at three different temperatures, viz., RT, 200 and 100K using maximum entropy method. The 2-dimensional charge density was mapped in order to visualize the bonding and the piling up of charges from the cation and anion sites. The 1-dimensional profiles were plotted along the crystallographic directions of [100], [110] and [111] to quantify the results. The results reveal that, a predominant ionic along with covalent bonding exists in the material (Saravanan et al., 2005).

A work on the investigation of the bonding in some oxides viz., MgO, CaO, SrO and BaO, uses maximum entropy method to evaluate the electron density distributions (Israel et al., 2003). Deformation of oxygen was observed in all four systems, which gives excess of charge to the nearest neighbors. It can also be observed that, the deformation is less in MgO, and is very strong as the atomic number of the cation increases. The ionicity in the systems was also calculated. The interaction between the ions in MgO was found to be purely ionic, whereas, interaction leading to a covalent nature was found in the other three systems. An error analysis was also done in the above mentioned work, in order to show that, the quality of the data and the obtained results are reliable to understand the behavior of the system.

The electron density distribution analysis is helpful in correlating the properties and atomic arrangement of a material. The valence charge distribution of germanium at two different temperatures 296 and 200 K has been explained using the maximum entropy method and multipole model. The quantum chemical parameters related to bonding have been determined using both MEM and multipole methods (Israel et al., 2009).

In a work done by Syed Ali et al., (2010), the maximum entropy method was used to calculate the electron density of nano sized magnetic semiconductor Co^{2+}:ZnO from the X-ray intensity data. The charge densities of Co^{2+}:ZnO were analyzed for various temperatures and correlated to the magnetic properties. Between 600°C and 700°C, the ferromagnetic property was found to switch on and then switch off. The charge ordering causing the ferromagnetic property in Co^{2+}:ZnO was also discussed (Syed Ali et al., 2010).

Mn doped CeO_2 nanosystems were prepared by chemical co-precipitation technique and were analyzed for the charge density properties which were correlated with their magnetic properties. It was found that, the magnetic ordering allows transition from diamagnetism to ferromagnetism with the increasing concentration of Mn^{2+} ions (Saravanakumar et al., 2014).

Apart from the quoted works, there are many research works in which the maximum entropy method and other methods are employed, in order to analyze the charge density distribution of materials and to correlate the properties of materials. These results inspire us to carry out the charge density distribution analysis for the materials we chose to investigate.

References

[1]   Attard A.E., Azaroff L.V., Journal of Applied Physics, 34 (4), 774 (1963). https://doi.org/10.1063/1.1729533

[2]   Collins D.M. Nature, 298, 49 (1982). https://doi.org/10.1038/298049a0

[3]   Coppens P., "X-Ray charge densities and chemical bonding", International Union of Crystallography, Oxford University Press, New York, 1997.

[4] Dawson B., Proceedings of the Royal Society, A, 298, 264 (1967a). https://doi.org/10.1098/rspa.1967.0103

[5] Dawson B., Proceedings of the Royal Society, A, 298, 395 (1967b). https://doi.org/10.1098/rspa.1967.0111

[6] Dawson B., Proceedings of the Royal Society, A, 298, 379 (1967c). https://doi.org/10.1098/rspa.1967.0110

[7] Flaig R., Koritsanszky T., Dittrich B., Wagner A., Luger P., J. Am. Chem. Soc. 124, 3407 (2002). https://doi.org/10.1021/ja011492y

[8] Gull S.F., Daniel G.J., Nature, 272, 686 (1978). https://doi.org/10.1038/272686a0

[9] Israel S., Saravanan R., Srinivasan N., Mohanlal S.K., Journal of Physics and Chemistry of Solids, 64, 879 (2003). https://doi.org/10.1016/S0022-3697(02)00434-1

[10] Israel S., Syed Ali K.S., Sheeba R.A.J.R., Saravanan R., Journal of Phys. Chem. Solids, 70, 1185 (2009). https://doi.org/10.1016/j.jpcs.2009.07.002

[11] Jeong I.K., Proffen T., Mohiuddin-Jacobs F., Billinge S.J.L., J. Phys. Chem. A 103, 921 (1998). https://doi.org/10.1021/jp9836978

[12] Macchi P., Sironi A., Coord Chem Rev. 238, 383 (2003). https://doi.org/10.1016/S0010-8545(02)00252-7

[13] Proffen T., Billinge S.J.L., Journal of Applied Crystallography, 32, 572 (1999). https://doi.org/10.1107/S0021889899003532

[14] Rietveld H.M., Journal of Applied Crystallography, 2, 65 (1969). https://doi.org/10.1107/S0021889869006558

[15] Saravanan R., Israel S., Physica B, 352, 220 (2004). https://doi.org/10.1016/j.physb.2004.07.014

[16] Saravanan R., Israel S., Rajaram R.K., Physica B, 363, 166 (2005). https://doi.org/10.1016/j.physb.2005.03.018

[17] Saravanakumar S., Sasikumar S., Israel S., Pradhiba G.R., Saravanan R., Materials Science in Semiconductor Processing, 17, 186 (2014). https://doi.org/10.1016/j.mssp.2013.10.002

[18] Syed Ali K.S., Saravanan R., Israel S., Açıkgöz M., Arda L., Physica B, 405, 1763 (2010). https://doi.org/10.1016/j.physb.2010.01.036

Dilute Magnetic Semiconducting (DMS) Materials, R. Saravanan   Materials Research Forum LLC
Materials Research Foundations **35** (2018)   doi: http://dx.doi.org/10.21741/9781945291777

Chapter 3

Sample Preparation and Structural Analysis

Abstract

Chapter 3 deals with the method of preparation of the samples of the proposed diluted magnetic semiconductor materials and the structural analysis of the prepared DMS materials. The materials prepared using the high temperature melt growth technique are as follows: (i) $Ge_{1-x}Mn_x$ (x = 0, 0.04, 0.06, 0.10); (ii) $Ge_{1-x}V_x$ (x = 0.03, 0.06, 0.09); (iii) $Ge_{1-x}Co_x$ (x = 0.03, 0.06, 0.09). The materials prepared using the ball milling technique are as follows: (i) $Si_{1-x}Mn_x$ (x = 0.02) ; sample milled for 100h and 200h (ii) $Si_{1-x}Ni_x$ (x = 0, 0.03, 0.06, 0.09, 0.12). The morphology of the proposed materials is studied and presented in this chapter. The structural analysis of the proposed materials is done and the results are presented in this chapter.

Keywords

Melt Growth, Ball Milling, GeMn, GeV, GeCo, SiMn, SiNi, Morphology

Contents

3.1 Sample preparation

Among the various techniques available for the preparation of diluted magnetic semiconductors, high temperature melt growth and ball milling have been used widely. The high temperature melt technique has been adopted in the laboratory environment for the materials having a melting point of less than 1100°C (the maximum temperature up to which the evacuated quartz tubes can withstand). Diluted magnetic semiconductors having a host material like germanium (Ge) and having the melting point of 938°C (Haynes, 2014) are processed in this method. The method of ball milling has been adopted for materials having higher melting point. Silicon based diluted magnetic semiconductor materials in which, silicon is chosen as the host material (melting point is 1414°C) (Haynes, 2014), are prepared using the ball milling technique.

In this work the techniques of high temperature melt growth and ball milling have been adopted for the preparation of germanium and silicon based diluted magnetic semiconductors.

The DMS materials grown using the high temperature melt technique are:

$Ge_{1-x}Mn_x$ (x = 0, 0.04, 0.06, 0.10)

$Ge_{1-x}V_x$ (x = 0.03, 0.06, 0.09)

$Ge_{1-x}Co_x$ (x = 0.03, 0.06, 0.09)

The DMS materials prepared using the ball milling technique are:

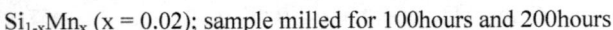

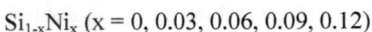

$Si_{1-x}Mn_x$ (x = 0.02); sample milled for 100hours and 200hours

$Si_{1-x}Ni_x$ (x = 0, 0.03, 0.06, 0.09, 0.12)

Materials Research Forum LLC
doi: http://dx.doi.org/10.21741/9781945291777

3.1.1 Preparation of the samples using the melt technique

In this work, the samples of $Ge_{1-x}M_x$ (M = Mn, Co, V) have been prepared using the melt technique. The atomic weights of the chemicals used are given in table 3.1. Initially, high purity powders of germanium (Ge, 99.99%) and manganese (Mn, 99.99%) were taken according to the stoichiometric ratio, for preparing three different compositions of $Ge_{1-x}Mn_x$ (x = 0.04, 0.06, 0.10) as given in table 3.2. The same stoichiometric calculations were carried out for the preparation of the diluted magnetic semiconductors $Ge_{1-x}V_x$(x = 0.03, 0.06, 0.09) and $Ge_{1-x}Co_x$ (x = 0.03, 0.06, 0.09). The weighed samples of the elemental powders were loaded in quartz tubes. The quartz tubes containing the samples were evacuated to a pressure of about10^{-6} Torr and vacuum sealed. The vacuum pumping system used for evacuating the quartz tubes containing the samples is shown in figure 3.1. The sample of undoped Ge was also prepared along with the other DMS samples. The evacuated quartz tubes containing the samples were placed inside a microprocessor controlled high temperature furnace shown in figure 3.2. The temperature of the furnace was increased from 30°C (room temperature) to 900°C, at a heating rate of 50°C/hr. When the temperature of 900°C was reached, the heating rate was reduced to 20°C/hr, until a temperature of 1050°C was reached. At this temperature, the melt was kept for soaking for 24 hours. Then, the melt was brought to the temperature of 950°C, at a cooling rate of 25°C/hr. At this temperature, the melt was again left for soaking for 6 hours. Then, it was cooled slowly at the rate of 5°C/hr, till the room temperature.

For the preparation of $Ge_{1-x}V_x$ (x = 0.03, 0.06, 0.09) and $Ge_{1-x}Co_x$ (x = 0.03, 0.06, 0.09), elemental powders of the host Ge and the dopants V and Co of high purity (99.99%) were weighed as per stoichiometric requirement as given in tables 3.3 and 3.4. The temperature of the furnace was increased from 30°C to 1050°C, at the heating rate of 20°C/hr. The soaking time given for the melt was 24 hours. Then, the melt was cooled slowly at the rate of 10°C/hr. After attaining a temperature of 950°C, the melt was again given a soaking time of 5 hours. Then, it was cooled to 30°C at the rate of 10°C/hr.

The melt grown samples of $Ge_{1-x}Mn_x$ (x = 0, 0.04, 0.06, 0.10), $Ge_{1-x}V_x$ (x = 0.03, 0.06, 0.09) and $Ge_{1-x}Co_x$ (x = 0.03, 0.06, 0.09) are presented as figure 3.3, figure 3.4 and figure 3.5 respectively.

Dilute Magnetic Semiconducting (DMS) Materials, R. Saravanan   Materials Research Forum LLC
Materials Research Foundations **35** (2018)   doi: http://dx.doi.org/10.21741/9781945291777

Table 3.1 Atomic weights of chemicals used.

Chemicals	Atomic weight (a.u.)
Silicon (Si)	28.086
Germanium (Ge)	72.61
Manganese (Mn)	54.94
Vanadium (V)	50.94
Cobalt (Co)	58.93
Nickel (Ni)	58.69

Table 3.2 Quantities of starting materials and growth conditions used to prepare $Ge_{1-x}Mn_x$.

Concent-ration (x)	Ge (gm)	Co (gm)	T_0 (°C)	T_F (°C)	ST_1 (°C)	St_1 (hr)	ST_2 (°C)	St_2 (hr)	HR_1 (°C/hr)	CR_1 (°C/hr)	HR_2 (°C/hr)	CR_2 (°C/hr)
0.04	1	0.042	30	1050	1050	24	950	6	50	25	20	5
0.06	1	0.064	30	1050	1050	24	950	6	50	25	20	5
0.10	1	0.111	30	1050	1050	24	950	6	50	25	20	5

T_0 – Initial temperature
T_F – Final temperature
ST_1 – Soaking temperature 1
St_1 – Soaking time 1
ST_2 – Soaking temperature 2
St_2 – Soaking time 2
HR_1 – Heating rate 1
CR_1 – Cooling rate 1
HR_2 – Heating rate 2
CR_2 – Cooling rate 2

Table 3.3 Quantities of starting materials and growth conditions used to prepare $Ge_{1-x}V_x$.

Concent -ration (x)	Ge (gm)	V (gm)	T_0 (°C)	T_F (°C)	ST_1 (°C)	St_1 (hr)	ST_2 (°C)	St_2 (hr)	HR (°C/ hr)	CR (°C/ hr)
0.04	1	0.042	30	1050	1050	24	950	5	20	25
0.06	1	0.064	30	1050	1050	24	950	5	20	25
0.10	1	0.111	30	1050	1050	24	950	5	20	25

T_0 – Initial temperature
T_F – Final temperature
ST_1 – Soaking temperature 1
St_1 – Soaking time 1
ST_2 – Soaking temperature 2
St_2 – Soaking time 2
HR – Heating rate
CR – Cooling rate

Table 3.4 Quantities of starting materials and growth conditions used to prepare $Ge_{1-x}Co_x$.

Concent -ration (x)	Ge (gm)	Co (gm)	T_0 (°C)	T_F (°C)	ST_1 (°C)	St_1 (hr)	ST_2 (°C)	St_2 (hr)	HR (°C/ hr)	CR (°C/ hr)
0.04	1	0.042	30	1050	1050	24	950	5	20	25
0.06	1	0.064	30	1050	1050	24	950	5	20	25
0.10	1	0.111	30	1050	1050	24	950	5	20	25

T_0 – Initial temperature
T_F – Final temperature
ST_1 – Soaking temperature 1
St_1 – Soaking time 1
ST_2 – Soaking temperature 2
St_2 – Soaking time 2
HR – Heating rate
CR – Cooling rate

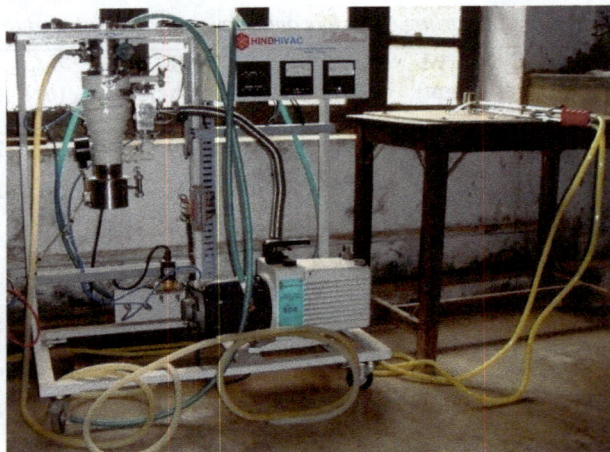

Figure 3.1 Vacuum pumping system used to evacuate the quartz tubes containing the samples (Hind Hivac Make; Model VS – 65D; 10^{-6} Torr).

Figure 3.2 Microprocessor controlled high temperature furnace used for melt growth of the DMS samples (Technico Make; Max. temp: 1200°C).

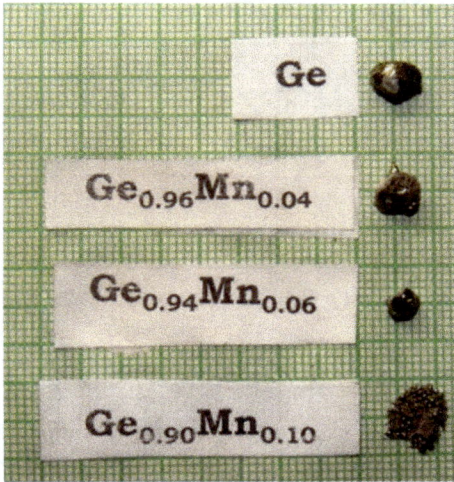

Figure 3.3 Melt grown samples of $Ge_{1-x}Mn_x$.

Figure 3.4 Melt grown samples of $Ge_{1-x}V_x$.

Figure 3.5 Melt grown samples of $Ge_{1-x}Co_x$.

3.1.2 Preparation of $Si_{1-x}Mn_x$ using the ball milling technique

The ball milling technique is adopted for the preparation of the silicon based diluted magnetic semiconductors since, the melting point of the host element silicon (Si) is high (1414°C) (Haynes, 2014). High purity (99.99%) elemental powders of silicon (Si) and manganese (Mn) were weighed as per the stoichiometric ratio for the preparation of $Si_{0.98}Mn_{0.02}$ as given in table 3.5. The elemental powders of the host and dopant materials were loaded in a stainless steel cylinder of the laboratory ball mill (figures 3.6(a) and (b)), along with the stainless steel balls (figure 3.7) of three different diameters 6 mm, 8 mm and 10 mm. The sample was milled for 100 hours and 200 hours with an average rpm of 200. The direction of rotation was reversed every hour. A pause time of 1 minute was given for every 15 minutes in order to avoid the generation of heat due to friction. Since the laboratory ball milled used for the preparation of Si:Mn DMS material was a low energy ball mill, the samples were milled up to 200 h to ensure that the doping process is successful. The ball milled powders were then taken and analyzed for its structural properties using X-ray diffraction data.

3.1.3 Preparation of $Si_{1-x}Ni_x$ using the ball milling technique

High purity (99.99%) powders of silicon (Si) and nickel (Ni) were weighed in appropriate quantities to get the stoichiometric ratio of the samples of $Si_{1-x}Ni_x$ (x = 0.03, 0.06, 0.09, 0.12) as given in table 3.6. The weighed powder samples of the host Si and dopant Ni for $Si_{1-x}Ni_x$ (x = 0.03, 0.06, 0.09, 0.12) were loaded in the stainless steel container of the high energy ball mill (Retsch Make, Model PM100), and milled for 4 hours. The milling media were four stainless steel balls each with the diameter of 1 cm and a mass of 4 gms.

The milling speed was 250 rpm. All the four samples were milled under the same environmental conditions.

Table 3.5 Quantities of starting materials used to prepare $Si_{1-x}Mn_x$.

Concentration (x)	Si (gm)	Mn (gm)	Milling time (h)	Milling speed (rpm)	Mass of milling media (gm)
0.02	2	0.04	100	200	4
0.02	2	0.04	200	200	4

Table 3.6 Quantities of starting materials used to prepare $Si_{1-x}Ni_x$.

Concentration (x)	Si (gm)	Ni (gm)	Milling time (h)	Milling speed (rpm)	Mass of milling media (gm)
0.03	2.72	0.084	4	250	4
0.06	2.72	0.168	4	250	4
0.09	2.72	0.252	4	250	4
0.12	2.72	0.336	4	250	4

(a)   **(b)**

Figure 3.6 Laboratory ball mill (a) side view (b) front view (With stainless steel cylinder for grinding; Direction reversible; 250 rpm).

Dilute Magnetic Semiconducting (DMS) Materials, R. Saravanan   Materials Research Forum LLC
Materials Research Foundations **35** (2018)   doi: http://dx.doi.org/10.21741/9781945291777

Figure 3.7 Milling media (Stainless steel balls) used to grind the elemental powders.

3.2   Morphological studies

The morphological study has been carried out by recording the micrographs using scanning electron microscopy (SEM) which is helpful in visualizing the surface structure of the prepared DMS materials.

3.2.1   SEM analysis of $Ge_{1-x}V_x$

The scanning electron microscopy (SEM) micrographs recorded for the melt grown samples of $Ge_{1-x}V_x$ (x = 0.03, 0.06, 0.09) at the Sophisticated Test and Instrumentation Centre (STIC), Cochin University of Science and Technology, Kochi, Kerala, India, using the SEM instrument model JSM-6390 were analyzed to understand the morphological nature of $Ge_{1-x}V_x$. The SEM micrographs of the samples of $Ge_{0.97}V_{0.03}$, $Ge_{0.94}V_{0.06}$ and $Ge_{0.91}V_{0.09}$ are shown in figures 3.8 (a) to (c).

The grains of pure germanium are visible in the SEM micrograph. When the concentration of the dopant is x = 0.03, the grain boundaries lead to the termination of long order and meager crystalline nature. When the dopant concentration is increased to x = 0.06, the crystalline nature of the sample is improved. When the concentration of the dopant is x = 0.09, non-uniform grain concentration is observed. This leads to the dislocation in the system, which could possibly affect the magnetic properties of the system.

3.2.2   SEM analysis of $Ge_{1-x}Co_x$

The SEM analysis of grown samples of $Ge_{1-x}Co_x$ (x = 0.03, 0.06, 0.09) were carried out in order to know about the morphological pattern of the samples. The SEM micrographs for the $Ge_{1-x}Co_x$ samples were recorded at the Sophisticated Test and Instrumentation Centre (STIC), Cochin University of Science and Technology, Kochi, Kerala, India, using the SEM instrument of model JSM-6390. The SEM micrographs of the samples of $Ge_{0.97}Co_{0.03}$, $Ge_{0.94}Co_{0.06}$ and $Ge_{0.91}Co_{0.09}$ are shown in figures 3.9 (a) to (c).

From the SEM micrographs, it is clear that, the sample having the concentration of $x = 0.03$ has crystalline nature. When the dopant concentration is increased to $x = 0.06$, the sample has a non-uniform grain concentration. When the dopant concentration is further increased to $x = 0.09$, the crystalline nature of the sample is well pronounced. The incorporation of the Co atoms in the Ge lattice is responsible for the structural changes, and these changes affect the magnetic nature of the $Ge_{1-x}Co_x$ samples.

3.2.3 SEM analysis of $Si_{1-x}Mn_x$

The SEM analysis of the ball milled samples of $Si_{0.98}Mn_{0.02}(100h)$ and $Si_{0.98}Mn_{0.02}(200h)$ was carried out to understand the effects on the morphology of the samples due to the milling process. The SEM micrographs of the samples of $Si_{1-x}Mn_x$ were recorded using JEOL JSM-6000 SEM instrument, at Karunya University, Coimbatore, Tamil Nadu, India. The SEM micrographs of the ball milled samples of $Si_{0.98}Mn_{0.02}(100h)$ and $Si_{0.98}Mn_{0.02}(200h)$ are shown in figures 3.10 (a) and (b) respectively.

It is obvious that, the process of ball milling decreases the grain size of the DMS samples of $Si_{1-x}Mn_x$. The mechanical alloying process has initiated contamination in the samples, which has been confirmed by the EDAX analysis. Elements such as chromium, nickel and iron which are the constituents of stainless steel, were found to have been mixed with prepared DMS Mn:Si. These impurities came from the stainless steel balls used as the milling media and the stainless steel container in which the samples were loaded for the mechanical alloying process. But, the contamination level was very low, and hence, the effect of the contaminations has been neglected.

3.2.4 SEM analysis of $Si_{1-x}Ni_x$

In order to understand the morphology of the ball milled samples of $Si_{1-x}Ni_x(x = 0.03, 0.06, 0.09, 0.12)$, SEM analysis has been carried out. The SEM micrographs were recorded using JEOL JSM 6390, at Sophisticated Test and Instrumentation Centre (STIC), Cochin, Kerela, India. The SEM micrographs of the samples of $Si_{0.97}Ni_{0.03}$, $Si_{0.94}Ni_{0.06}$, $Si_{0.91}Ni_{0.09}$ and $Si_{0.88}Ni_{0.12}$ are shown in figures 3.11(a) to (d).

The SEM micrographs of the samples show that, the size of the particles is of the order of micrometers. Also, uniform micro sized particles are observed throughout the samples. The agglomeration of the particles seems to have increased with the increase in the concentration of the dopant in the host lattice. The homogeneity of the sample has been verified in the SEM measurements, and hence, expected to have uniform experimental

Dilute Magnetic Semiconducting (DMS) Materials, R. Saravanan   Materials Research Forum LLC
Materials Research Foundations **35** (2018)   doi: http://dx.doi.org/10.21741/9781945291777

behavior. The mixed samples were verified for the stoichiometric contents using EDS elemental analysis, and found to have $x = 0.0278, 0.054, 0.081$ and 0.109 instead of the expected $x = 0.03, 0.06, 0.09$ and 0.12 respectively and the difference is well within experimental deviations.

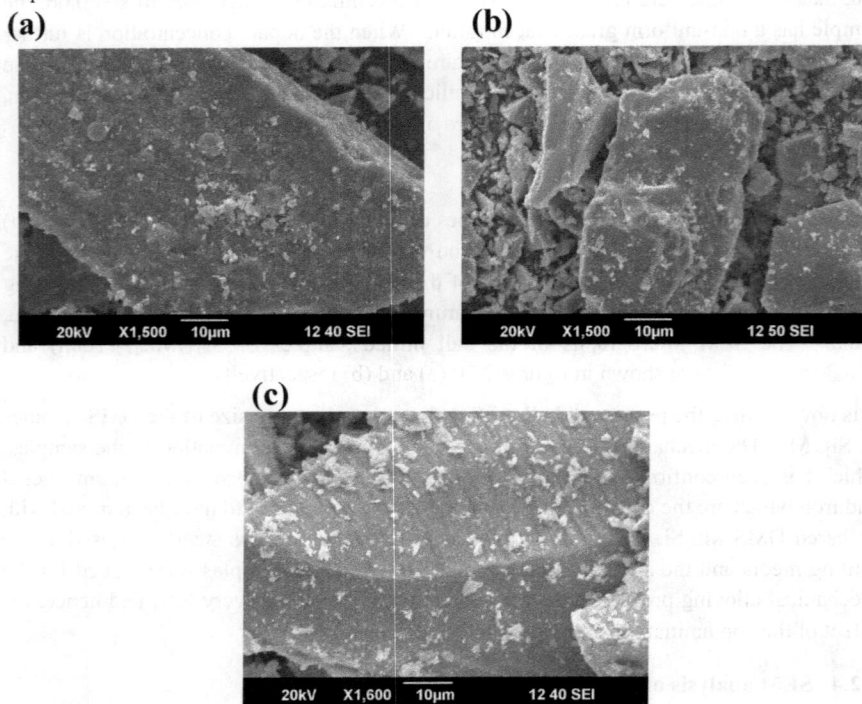

(a)   **(b)**

(c)

Figure 3.8 SEM micrographs of (a) $Ge_{0.97}V_{0.03}$ (b) $Ge_{0.94}V_{0.06}$ (c) $Ge_{0.91}V_{0.09}$.

(a)

(b)

(c)

Figure 3.9 SEM micrographs of (a) $Ge_{0.97}Co_{0.03}$ (b) $Ge_{0.94}Co_{0.06}$ (c) $Ge_{0.91}Co_{0.09}$.

(a)

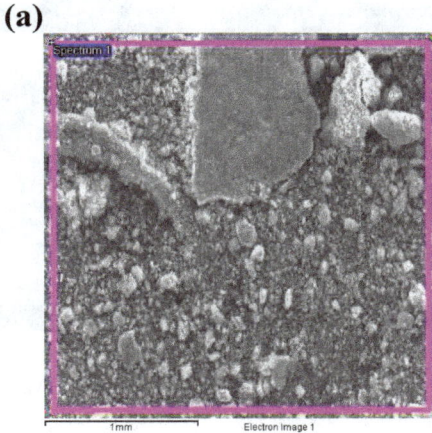

(b)

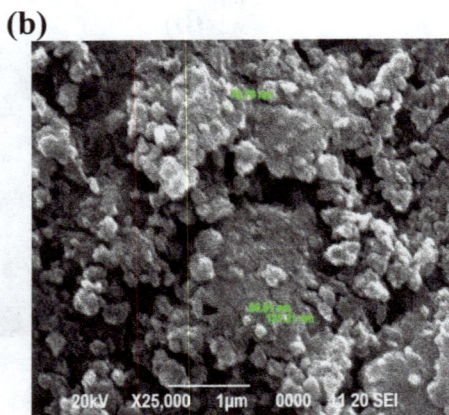

Figure 3.10 SEM micrographs of (a) $Si_{0.98}Mn_{0.02}(100h)$ (b) $Si_{0.98}Mn_{0.02}(200h)$.

(a)   **(b)**

(c)   **(d)**

Figure 3.11 SEM micrographs of (a) $Si_{0.97}Ni_{0.03}$ (b) $Si_{0.94}Ni_{0.06}$ (c) $Si_{0.91}Ni_{0.09}$ (d) $Si_{0.88}Ni_{0.12}$.

3.3 Structural properties

The structural properties of the prepared diluted magnetic semiconductor (DMS) materials have been analyzed from the X-ray diffraction experiment, employing the Rietveld refinement technique (Rietveld, 1969). The average structures of the materials have been analyzed using the software JANA 2006 (Petříček et al., 2014) and the structure factors extracted from the observed powder X-ray diffraction data are used to analyze the structural properties of the prepared DMS materials.

Table 3.7 Structural parameters for $Ge_{1-x}Mn_x$ through refinement of powder XRD data.

Parameter	Ge	$Ge_{0.96}Mn_{0.04}$	$Ge_{0.94}Mn_{0.06}$	$Ge_{0.90}Mn_{0.10}$
a (Å)	5.6558(7)	5.6469(2)	5.6676(14)	5.6645 (10)
B_{iso} (Å^2)	1.0302(0)	1.1352(0)	0.8599(0)	1.2665(0)
F_{000}	256	254	253	250
R_{obs} (%)	4.19	4.20	3.09	4.78
wR_{obs} (%)	4.97	5.38	4.32	5.67
R_p(%)	4.86	8.21	9.38	5.60
wR_p(%)	6.48	11.25	12.70	7.45
GoF	0.49	0.64	0.69	0.41
σ $(\theta)_{222}$(%)	0.0275	0.0765	0.1138	0.0958
σ $(\theta)_{115/333}$(%)	0.0367	0.1032	0.1514	0.0846
σ $(\theta)_{404}$(%)	0.0864	0.1755	0.1493	0.1005
σ $(\theta)_{315}$(%)	0.0904	0.1992	0.1467	0.1102

σ– Percentage deviation in the Bragg angle of the corresponding Bragg reflection
a-Cell dimension
B_{iso}– Thermal parameter
R_{obs}-Reliability index for observed structure factors
wR_{obs}-Weighted reliability index for observed structure factors
R_p-Reliability index for profile
wR_p– Weighted reliability index for profile
F_{000}-Number of electrons in the unit cell
GoF-Goodness of fit

3.3.1 Structural analysis of $Ge_{1-x}Mn_x$ (x = 0, 0.04, 0.06, 0.10)

The melt grown samples of $Ge_{1-x}Mn_x$ (x = 0, 0.04, 0.06, 0.10) were ground into fine powders using agate mortar and pestle. The powder X-ray diffraction data of the samples of $Ge_{1-x}Mn_x$ were collected at National Institute for Interdisciplinary Science and

Technology (NIIST), Trivandrum, India, using XPERT PRO (Philips, Netherlands) powder diffractometer. The wavelength of the X-ray used was Cu $K_{\alpha1}$ wavelength of 1.54056Å, with the 2θ range from 10° to 120° and an interval of 0.02°. The Rietveld (Rietveld, 1969) refined X-ray diffraction profiles of the samples of $Ge_{1-x}Mn_x$ (x = 0, 0.04, 0.06, 0.10) are shown in figures 3.12 (a) to (d). The values of the refined parameters and the reliability indices are given in table 3.7. The space group of the host system (Ge) is $Fd\bar{3}m$. The total number of charges in the unit cell (F_{000}) is found to decrease with the increase in the dopant concentration, which ensures the incorporation of the dopant Mn atom in the host Ge lattice. The lattice is found to be compressed when x = 0.04, and expanded when x = 0.06. An increase in intensity of the peaks of (202), (313) and (224) is observed, when the doping concentration of Mn is increased. This increase in intensity confirms that, the increase of Mn dilution in the host system is linear with doping. When the transition metal atom Mn is introduced in the host semiconductor Ge, the d orbital of Mn atom and the p orbital of Ge atom gets hybridized that causes a local magnetic moment due to Mn atom in the Ge matrix (Weng et al., 2005). This magnetic moment in the host matrix of Ge increases the mixing of d orbital of Mn and p orbital of Ge. This results in the compression of the host Ge lattice, that further affects the structural and lattice parameters of the Ge:Mn system. The percentage deviation (σ) between the calculated and observed Bragg angles is compared for $Ge_{1-x}Mn_x$ (0, 0.04, 0.06, 0.10), using the formula $\{[(\theta_{obs}-\theta_{cal})/\theta_{cal}] \times 100\}$ and given in the table 3.7.

(a)

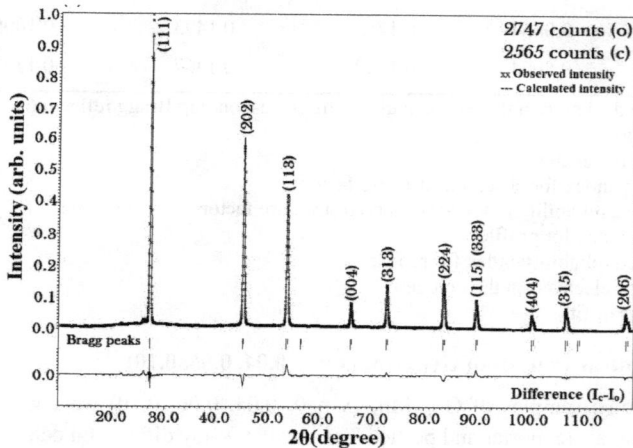

Figure 3.12 (a) Fitted powder XRD profile for Ge.

(b)

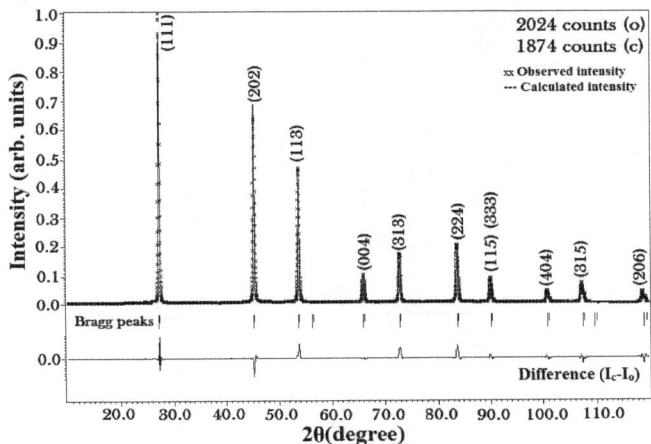

Figure 3.12 (b) Fitted powder XRD profile for $Ge_{0.96}Mn_{0.04}$.

(c)

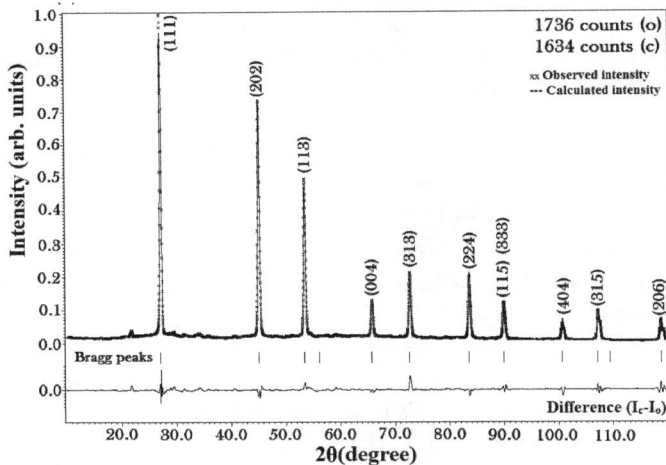

Figure 3.12 (c) Fitted powder XRD profile for $Ge_{0.94}Mn_{0.06}$.

(d)

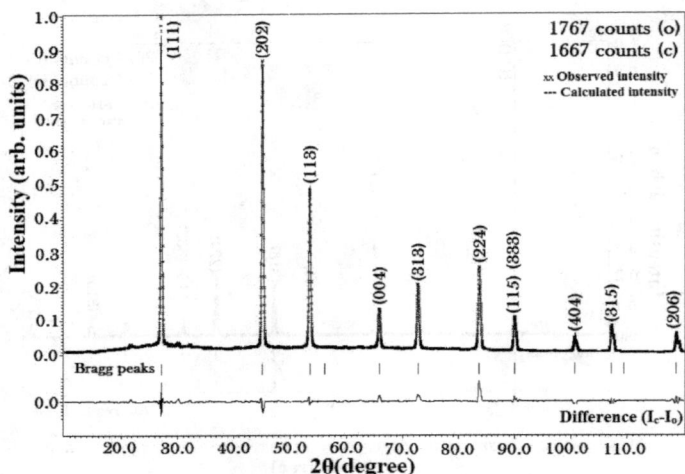

Figure 3.12 (d) Fitted powder XRD profile for $Ge_{0.90}Mn_{0.10}$.

3.3.2 Structural analysis of $Ge_{1-x}V_x$ (x = 0.03, 0.06, 0.09)

The melt grown polycrystalline samples of $Ge_{1-x}V_x$ (x = 0.03, 0.06 and 0.09) were ground using agate mortar and pestle, and made into fine powders. The powder X-ray diffraction data for all the three samples were recorded at Sophisticated Test and Instrumentation Centre (STIC), Cochin, Kerela, India, using Anton Paar, TTK 450 powder X-ray diffractometer. The $CuK_{\alpha 1}$ radiation of wavelength 1.54056Å was used with the range of 2θ from 10° to 120° in steps of 0.02°. The observed powder X-ray intensity profiles for the samples are shown in figure 3.13. The observed X-ray intensity patterns match with JCPDS (Joint Committee Powder Diffraction Standards-PDF No. #04-0545) database.

By observing the obtained experimental powder X-ray diffraction pattern, it is found that, there is no sign of additional peaks in the observed X-ray intensity pattern apart from those of the prepared system of $Ge_{1-x}V_x$. This confirms that, a nominal concentration of dopant vanadium atoms is incorporated at the substitutional positions in the host Ge lattice. It can be noted that, the intensity of the diffraction pattern decreases with the increase of dopant concentration. This is because, the dopant has a lower atomic number than the host [Z(V) = 23; Z(Ge) = 32]. The powder X-ray data sets were refined to attain the accurate structure factors from which, the charge densities are derived. Rietveld (Rietveld, 1969) refinement method is followed using the software JANA 2006 (Petříček et al., 2014) and the refined structural parameters are given in table 3.8. The refined profiles of the samples of $Ge_{1-x}V_x$ are given in figures 3.14 (a) to (c).

Table 3.8 Structural parameters for $Ge_{1-x}V_x$ through refinement of powder XRD data.

Parameter	$Ge_{0.97}V_{0.03}$	$Ge_{0.94}V_{0.06}$	$Ge_{0.91}V_{0.09}$
a (Å)	5.6543(2)	5.6558(5)	5.6585(4)
B_{iso} (Å^2)	0.4625(5)	0.5114(3)	0.9919(7)
F_{000}	253.84	251.68	249.52
R_{obs} (%)	2.70	2.09	3.45
wR_{obs} (%)	3.54	2.70	5.32
R_p(%)	3.85	3.40	4.00
wR_p(%)	5.46	4.65	6.46
Density(g cm^{-3})	5.2839(1)	5.2329(0)	5.1780(1)
GoF	0.92	0.81	1.08

a-Cell dimension
B_{iso}– Thermal parameter
R_{obs}-Reliability index for observed structure factors
wR_{obs}-Weighted reliability index for observed structure factors
R_p-Reliability index for profile
wR_p– Weighted reliability index for profile
F_{000}-Number of electrons in the unit cell
GoF-Goodness of fit

The decrease in the total number of charges in the unit cell (F_{000}) (Table 3.8) with the increase in dopant concentration is due to the lower atomic number of the dopant than the host atom. This confirms the incorporation of the dopant atom V at the regular lattice sites of the host lattice Ge. A shift in the Bragg peaks is observed towards the lower angles, when the concentration of V is increased. This shift in the Bragg peaks is due to the distortion at the host lattice sites by the addition of the impurity. The lattice constant values of the samples increase with the increase in dopant concentration because of the higher atomic number of the host atom (Ge), than that of the dopant atom (V). The value of the Debye-Waller factor (B_{iso}) increases as the concentration of the dopant atoms is increased in the host Ge lattice, since, the atomic weight of V is smaller than the atomic weight of Ge. Hence, the values of the Debye-Waller factor increases from 0.4625Å^2 (3% of dopant concentration) to 0.5114Å^2 (6% of dopant concentration), and then to 0.9919Å^2 (9% of dopant concentration). The value of B_{iso} increases due to the presence of spin-density wave in the system, along with the spins arranged alternatively in the parallel and antiparallel direction. The presence of the spin-density wave eliminates the possibility of binding the atoms together tightly. Hence, the atoms are permitted to vibrate freely, with more root mean square amplitude than before. The observed and the calculated structure factors are tabulated in table 3.9 with the corresponding reflection planes. The lower atomic number of the dopant atom causes the decrease in the magnitude of the experimental structure factors.

Table 3.9 Observed and calculated structure factors for various Bragg reflections of $Ge_{1-x}V_x$.

h	k	l	$Ge_{0.97}V_{0.03}$			$Ge_{0.94}V_{0.06}$			$Ge_{0.91}V_{0.09}$		
			F_o	F_c	$\sigma(F_o)$	F_o	F_c	$\sigma(F_o)$	F_o	F_c	$\sigma(F_o)$
1	1	1	146.472	145.834	1.487	144.144	144.401	1.459	141.326	141.541	1.434
2	0	2	174.954	174.928	1.785	173.061	172.759	1.760	165.966	166.081	1.695
1	1	3	119.291	114.670	1.228	114.959	113.076	1.181	105.993	107.450	1.095
0	0	4	137.120	144.873	1.529	135.811	142.526	1.514	134.657	132.863	1.510
3	1	3	91.273	96.2230	0.986	93.7296	94.5406	1.005	100.284	87.1324	1.073
2	2	4	125.630	123.359	1.323	117.895	120.956	1.244	101.731	109.396	1.084
3	3	3	80.543	82.498	0.885	84.3147	80.798	0.925	75.263	72.259	0.845
1	1	5	80.543	82.498	0.885	84.2778	80.798	0.927	75.195	72.260	0.838
4	0	4	105.228	106.790	1.235	100.555	104.401	1.201	86.368	91.645	1.077
3	1	5	76.465	71.776	0.828	70.612	70.098	0.779	62.437	60.852	0.701
2	0	6	96.472	93.598	1.050	92.167	91.260	1.037	76.098	77.774	0.871

F_o-Observed structure factor
F_c-Calculated structure factor
$\sigma(F_o)$-Standard deviation error for observed structure factor

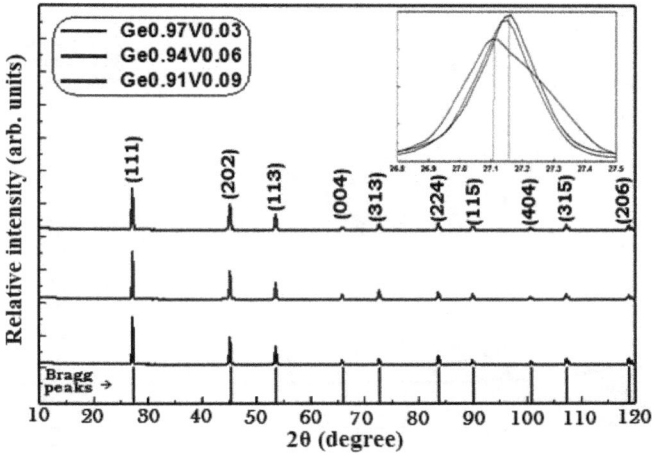

Figure 3.13 Observed powder X-ray diffractiogram of $Ge_{0.97}V_{0.03}$, $Ge_{0.94}V_{0.06}$ and $Ge_{0.91}V_{0.09}$.

(a)

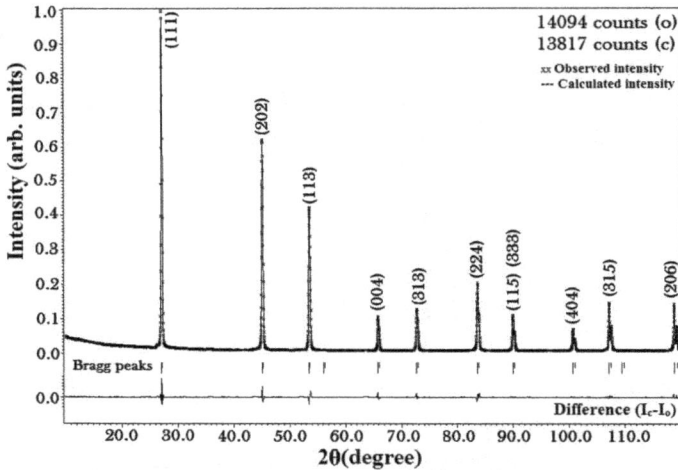

Figure 3.14 (a) Fitted powder XRD profile for $Ge_{0.97}V_{0.03}$.

(b)

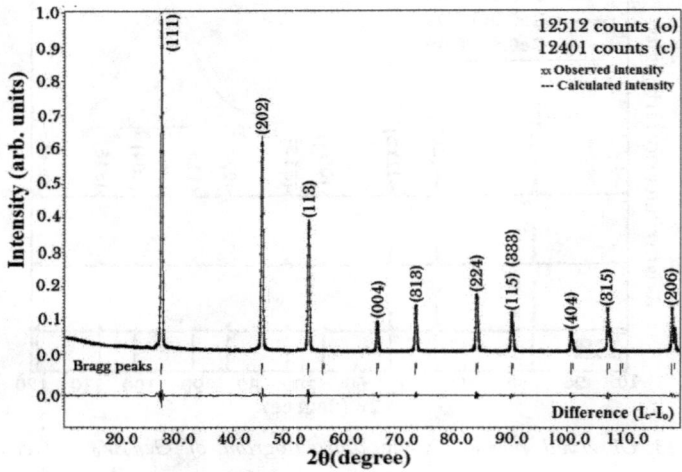

Figure 3.14 (b) Fitted powder XRD profile for $Ge_{0.94}V_{0.06}$.

(c)

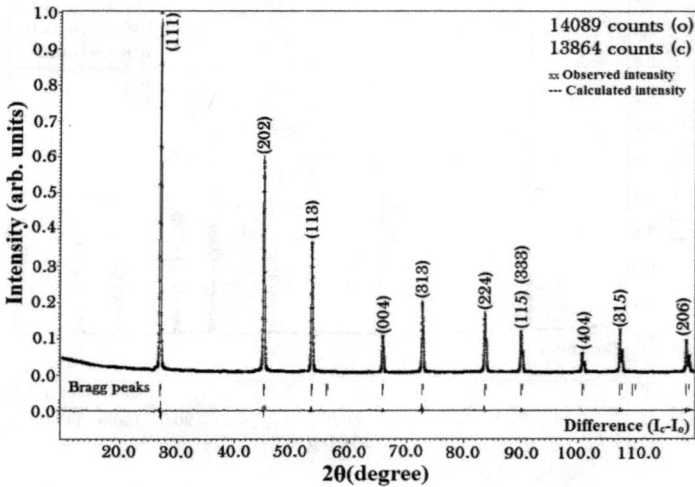

Figure 3.14 (c) Fitted powder XRD profile for $Ge_{0.91}V_{0.09}$.

3.3.3 Structural analysis of $Ge_{1-x}Co_x$ (x = 0.03, 0.06, 0.09)

The polycrystalline samples of $Ge_{0.97}Co_{0.03}$, $Ge_{0.94}Co_{0.06}$ and $Ge_{0.91}Co_{0.09}$ grown using the high temperature melt growth technique, were ground into powders using agate mortar and pestle. The powder X-ray diffraction data of the samples were recorded at the Sophisticated Test and Instrumentation Centre (STIC), Cochin, Kerela, India, using Anton Paar, TTK 450 powder X-ray diffractometer. The $CuK_{\alpha 1}$ wavelength of X-rays (1.54056 Å) was used in the range of 10° to 120° with the interval of 0.02°. The observed powder diffraction data sets are compared and shown in figure 3.15.

Table 3.10 Structural parameters for $Ge_{1-x}Co_x$ through refinement of powder XRD data.

Parameter	Ge	$Ge_{0.97}Co_{0.03}$	$Ge_{0.94}Co_{0.06}$	$Ge_{0.91}Co_{0.09}$
a (Å)	5.6558(7)	5.6565(3)	5.6530(2)	5.6561(5)
B_{iso} (Å^2)	1.0302(0)	0.5076	0.5000	0.5000
F_{000}	256	255	254	252
R_{obs} (%)	4.19	2.89	3.71	4.00
wR_{obs} (%)	4.97	3.66	4.70	4.60
R_p(%)	4.86	7.95	5.76	5.31
wR_p(%)	6.48	10.69	8.73	8.87
Density(g cm^{-3})	5.3285(7)	5.2963(3)	5.2762(2)	5.2373(5)
GoF	0.49	1.73	2.30	2.66
Phase fraction of GeCo (%)	0	0.83	0.85	0.89

a-Cell dimension
B_{iso}– Thermal parameter
R_{obs}-Reliability index for observed structure factors
wR_{obs}-Weighted reliability index for observed structure factors
R_p-Reliability index for profile
wR_p– Weighted reliability index for profile
F_{000}-Number of electrons in the unit cell
GoF-Goodness of fit

The experimental powder X-ray diffraction data sets of the DMS samples of $Ge_{1-x}Co_x$ were subjected to Rietveld (Rietveld, 1969) refinement, and the refined structural parameters for the system $Ge_{1-x}Co_x$ (x = 0.03, 0.06 and 0.09) are given in table 3.10. The Rietveld (Rietveld, 1969) refined profiles are given in figures 3.16 (a) to (c). The refinement of the crystal structure of $Ge_{1-x}Co_x$ was done by minimizing the difference

Dilute Magnetic Semiconducting (DMS) Materials, R. Saravanan   Materials Research Forum LLC
Materials Research Foundations **35** (2018)   doi: http://dx.doi.org/10.21741/9781945291777

between calculated and observed structure factors of the samples. It is found that, there is an increase in the value of the lattice constant (Table 3.10), when the dopant concentration is 3% compared to that of undoped germanium, whereas, the lattice constant decreases when the dopant concentration is 6%. When the dopant concentration is 9%, the value of the cell constant again increases (Table 3.10). It is also found that, the total charges in the unit cell (F_{000}) of the system of Ge:Co decrease, when the dopant concentration increases. It is clear from figure 3.15 that, a highly crystalline nature is found in the system from the peaks of the experimental diffraction pattern (small FWHM). At the same time, additional diffraction peaks of the monoclinic system of GeCo (JCPDS-PDF # 65-1223) are present in the observed powder X-ray diffraction pattern. The additional phase was found to be a small fraction. It is clear from the observed diffraction pattern (Figure 3.15) that, the intensity of the phase GeCo increases as the dopant concentration is increased, which evidences the inclusion of the monoclinic GeCo system (Mocking et al., 2012). The influence of both the systems, GeCo and Ge:Co, is found to be present in the net profile. The contributions are found from phase fractions, derived from the Rietveld (Rietveld, 1969) refinement as, 0.83% for 3% of dopant concentration, 0.85% for 6% of dopant concentration and 0.89% for 9% of dopant concentration.

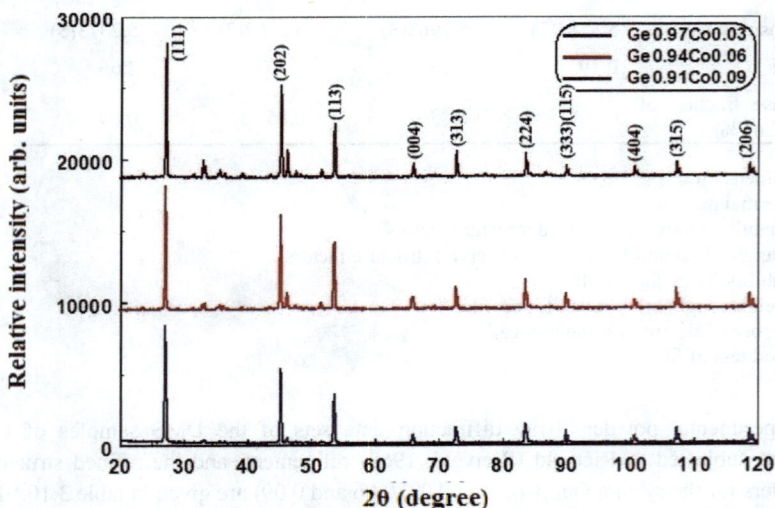

Figure 3.15 Observed powder X-ray diffractogram of $Ge_{0.97}Co_{0.03}$, $Ge_{0.94}Co_{0.06}$ and $Ge_{0.91}Co_{0.09}$.

(a)

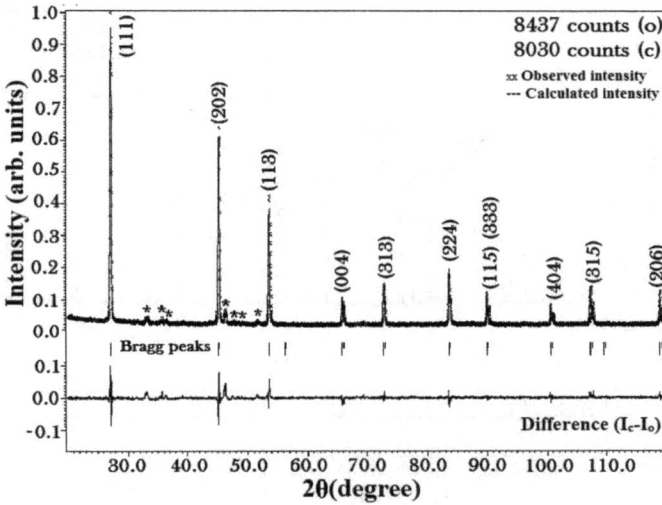

Figure 3.16 (a) Fitted powder XRD profile for $Ge_{0.97}Co_{0.03}$.

(b)

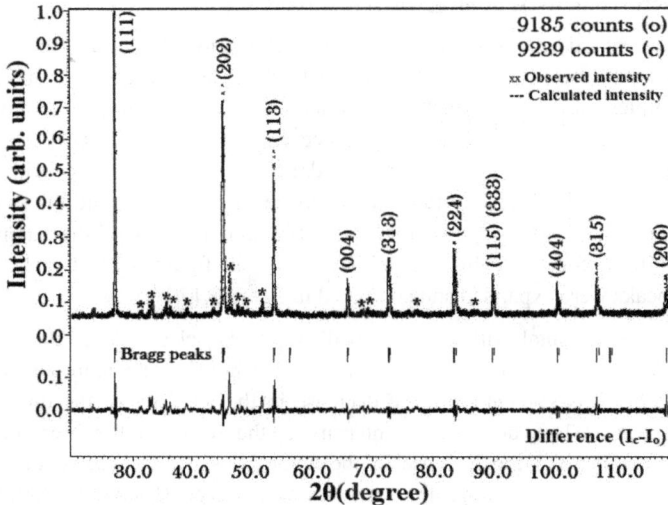

Figure 3.16 (b) Fitted powder XRD profile for $Ge_{0.94}Co_{0.06}$.

67

(c)

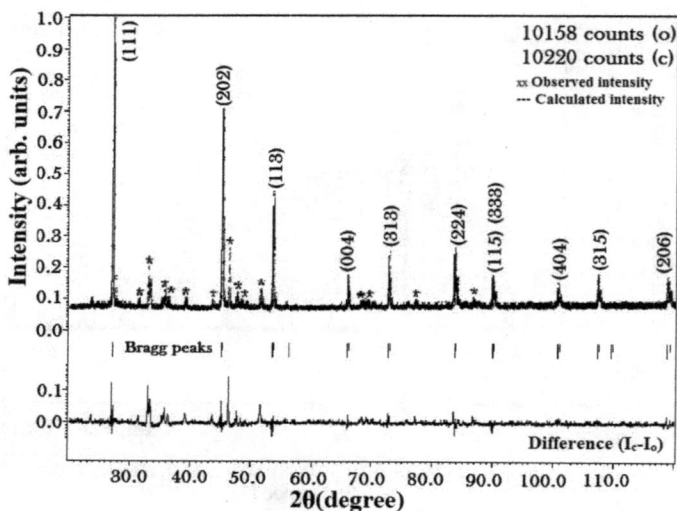

Figure 3.16 (c) Fitted powder XRD profile for $Ge_{0.91}Co_{0.09}$ (* - Monoclinic GeCo).

3.3.4 Structural analysis of $Si_{0.98}Mn_{0.02}$

The samples of $Si_{0.98}Mn_{0.02}$ ball milled for 100 and 200 hours are referred as $Si_{0.98}Mn_{0.02}$(100h) and $Si_{0.98}Mn_{0.02}$(200h) respectively. The powder X-ray diffraction data for the samples of Si, $Si_{0.98}Mn_{0.02}$(100h) and $Si_{0.98}Mn_{0.02}$(200h) were collected at, National Institute for Interdisciplinary Science and Technology (NIIST), Trivandrum, India, using XPERT PRO (Philips, Netherlands) powder diffractometer, with the $CuK_{\alpha1}$ wavelength (1.54056Å). The 2θ range was between 10° and 120° and the step size was 0.017°. The raw untreated powder X-ray diffraction profiles of the samples of Si, $Si_{0.98}Mn_{0.02}$(100h) and $Si_{0.98}Mn_{0.02}$(200h) are shown in figure 3.17(a). The first three prominent peaks were expanded and compared in figure 3.17(b).

From the experimental diffraction profiles, it is clear that, the sample of $Si_{0.98}Mn_{0.02}$(200h) is closing in to becoming amorphous. The amorphous nature of $Si_{0.98}Mn_{0.02}$ (200h) is evident from the decrease in the intensity of the observed X-ray diffraction peaks. This decrease in intensity is the result of the breaking down of crystalline domains and the increase in the amorphous nature, due to the process of milling. For refining the profiles of the samples, the Rietveld (Rietveld, 1969) refinement technique was used. The initial cell parameters were assumed as a = b = c = 5.43Å

(Wyckoff, 1963), and the symmetry group was $Fd\bar{3}m$. The refined profiles of the samples of Si, $Si_{0.98}Mn_{0.02}$(100h) and $Si_{0.98}Mn_{0.02}$(200h) are shown in figures 3.18 (a) to (c). The refined structural parameters of the samples and the corresponding reliability indices are given in table 3.11. The value of the cell dimension increases from that of the initial value of Si, which shows that, the dopant Mn is incorporated in the regular lattice site of Si (Table 3.11). Also, the mechanical infringement of Mn is more, as the time of milling increases. The atomic radius of Mn (2.731 Å)(Slater, 1964) is more than that of Si (2.352 Å) (Slater, 1964), and hence, the cell volume increases while the density decreases. The Debye-Waller factor (B_{iso}) for $Si_{0.98}Mn_{0.02}$(100h) decreases, when compared to that of pure Si, since, the heavier Mn atom is incorporated in the Si lattice. But, at the same time, B_{iso} of Si increases heavily for $Si_{0.98}Mn_{0.02}$(200h) since, the size of the grain/domain decreases, which accommodates the atoms in the surface of the grain and in the inner region, to have larger amplitude of vibration.

Table 3.11 Structural parameters for $Si_{1-x}Mn_x$ through refinement of powder XRD data.

Parameters	Si	$Si_{0.98}Mn_{0.02}$(100h)	$Si_{0.98}Mn_{0.02}$(200h)
a_0 (Å)	5.3462(9)	5.4432(6)	5.4507(7)
V(Å^3)	152.8040	161.4161	161.9423
ρ (gm/cc)	2.4409	2.3549	2.3472
B_{iso}(Å^2)	0.5500(0)	0.5350(0)	0.8679(0)
R_{obs}(%)	2.59	2.73	2.31
wR_{obs}(%)	2.28	1.93	2.68
GoF	0.46	0.48	0.51
D_v(nm)	15.74	10.44	8.88

a_0– Cell dimension
V-Volume of the unit cell
ρ-Density of the unit cell
B_{iso}– Thermal parameter
R_{obs}-Reliability index for observed structure factors
wR_{obs}-Weighted reliability index for observed structure factors
R_p-Reliability index for profile
F_{000}-Number of electrons in the unit cell
GoF-Goodness of fit
D_v-Size of the diffracting domain

Dilute Magnetic Semiconducting (DMS) Materials, R. Saravanan   Materials Research Forum LLC
Materials Research Foundations **35** (2018)   doi: http://dx.doi.org/10.21741/9781945291777

In the ball milled sample of $Si_{0.98}Mn_{0.02}$(100h), an additional phase of SiMn was found, and was refined to have cubic system with the space group of $P2_{1}3$, with the cell value of $a_0 = 4.5594$Å (JCPDS-PDF #42-1487). The secondary phase of SiMn has a signature peak (102) at 44.514° and has the composition 2.36%. The refinement of the profiles showed that, the grain size decreased as the milling time increased. Hence, there is lack of significant information on the secondary phase from the diffraction pattern of $Si_{0.98}Mn_{0.02}$(200h), except the enhancement of the peak (102). The grain size of the ball milled samples (D_v) was found using a code called GRAIN (Saravanan), in which, the Bragg angles and the corresponding pseudo-voigt FWHM from the X-ray diffraction data sets are given as inputs. The estimated size of the diffracting domains (using Debye-Scherrer formula) is given in table 3.11.

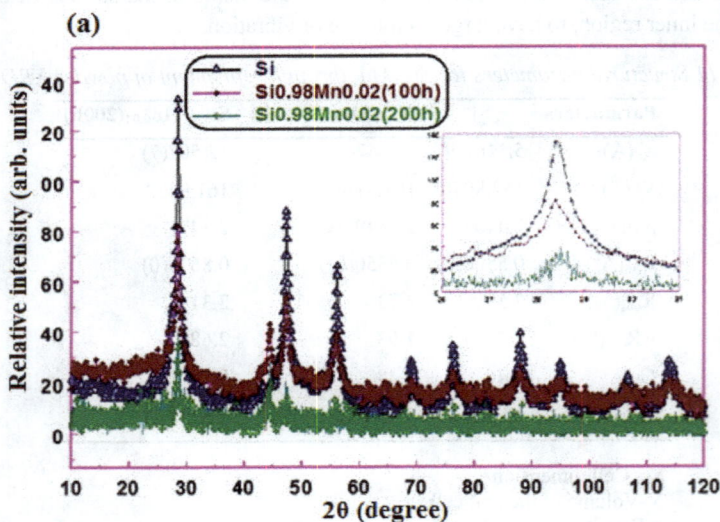

Figure 3.17 (a) Observed powder X-ray diffractogram of Si, $Si_{0.98}Mn_{0.02}$ (100h) and $Si_{0.98}Mn_{0.02}$ (200h).

(b)

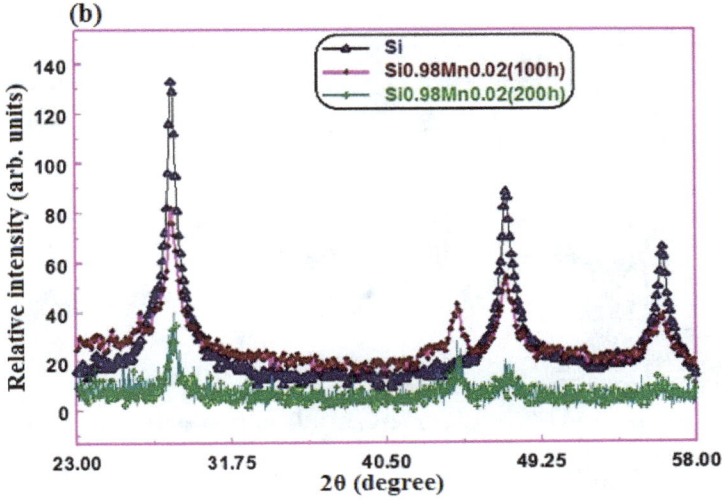

Figure 3.17 (b) Expansion of the first three prominent peaks in the observed powder X-ray diffractogram.

(a)

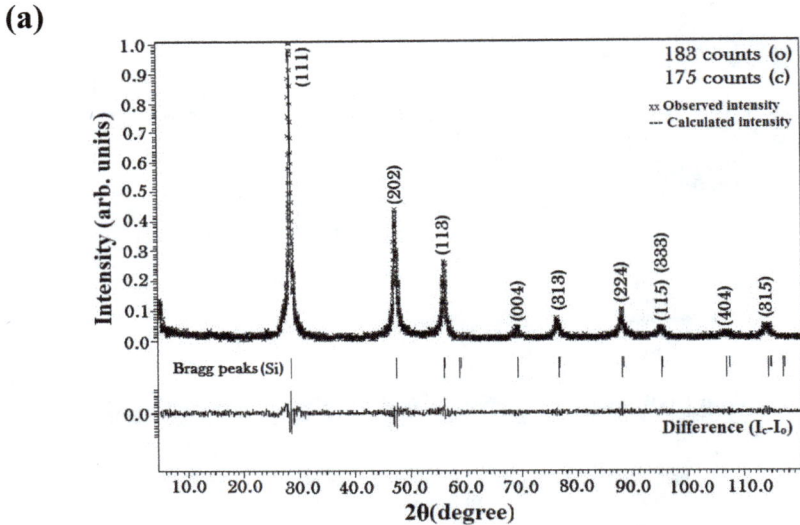

Figure 3.18 (a) Fitted powder XRD profile for Si.

(b)

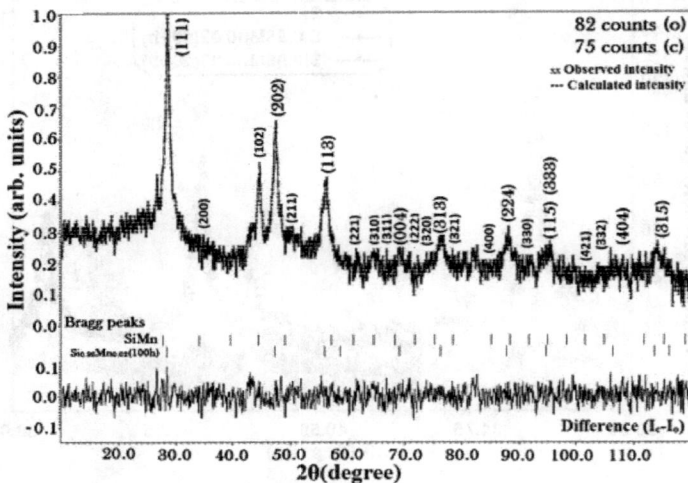

Figure 3.18 (b) Fitted powder XRD profile for $Si_{0.98}Mn_{0.02}$ (100h).

(c)

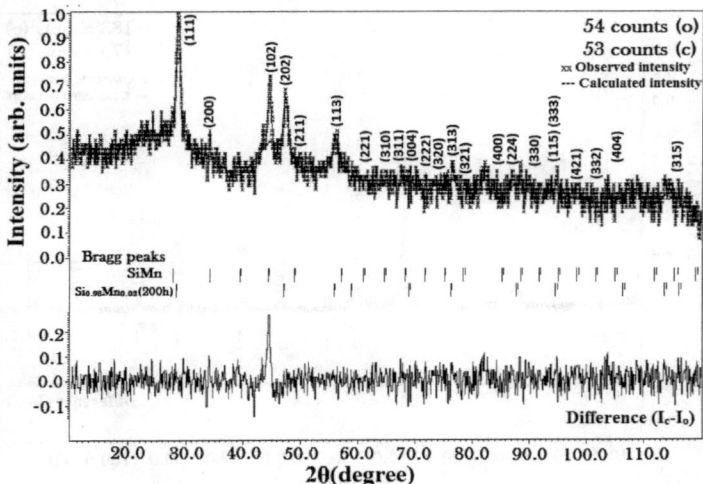

Figure 3.18 (c) Fitted powder XRD profile for $Si_{0.98}Mn_{0.02}$(200h).

3.3.5 Structural analysis of $Si_{1-x}Ni_x$ (x = 0, 0.03, 0.06, 0.09, 0.12)

The powder X-ray diffraction data for the powder samples of raw Si and the ball milled samples of $Si_{1-x}Ni_x$ (x = 0, 0.03, 0.06, 0.09, 0.12), were collected at the International Research Centre, Kalasalingam University, Krishnan Koil, Tamil Nadu, India, using Bruker (D8 Advance ECO XRCD Systems) powder X-ray diffractometer, using $CuK_{\alpha 1}$ wavelength (1.54056 Å) and 2θ range from 5° to 120° with an interval of 0.02°. The observed powder X-ray intensity profiles of the samples are presented in figure 3.19(a).

The structure factors of the crystal structures were derived from the observed powder X-ray diffraction data, and compared with that of the calculated ones, and refined till a minimum difference is obtained between them. The refined parameters and derived reliability indices are presented in table 3.12.

The comparison of the observed powder X-ray diffraction patterns of the samples with the Joint Committee Powder Diffraction Standards (JCPDS) (JCPDS-PDF #04-0850) showed phase pure system for x = 0, with diamond structure, having $Fd\bar{3}m$ as the space group. A small fraction of additional phase of pure Ni metal was found in all the doped samples, with FCC structure and $Fm\bar{3}m$ as the space group. A shift in the diffraction angles was observed in the observed powder X-ray diffraction profiles, when the dopant concentration was increased (Figure 3.19(b)). The value of the cell dimension of the unit cell was observed to increase with the increase in the dopant concentration (Table 3.12). The total number of electrons in the unit cell (F_{000}) is observed to increase with the increase in the dopant concentration (Table 3.12). Thus, the inclusion of Ni in the host lattice of Si is confirmed. The Rietveld refined profiles of the prepared samples of $Si_{1-x}Ni_x$ (x = 0, 0.03, 0.06, 0.09, 0.12) are presented in figures 3.20 (a) to (e). Ni cluster as an additional phase, was also observed in the prepared materials as additional peaks, due to the inherent low soluble nature of Ni in semiconductor host materials. The low solubility problem and less impurity interaction of Ni when doped with silicon where the bigger size of Ni atom were reported to be unfavorable for host Si matrix (Collver, 1977). It was also observed that, the semiconductor host with Ni will have the solubility of about 10^{17} cm^{-3}, resulting in an average inter impurity (Ni) separation of 130 Å. This amount of inter impurity (Ni) separation of around 60 unit cells, or in other words, 3.33% of occupancy of Ni atom, is possible in a single unit cell. Thus, the solubility level was found to be low as per binding energy estimation. However, our attempt on dissolving Ni into the host lattice sites more than that established by Collver (1977) proved futile, as the preparation of the mixture is done using the ball milling technique. Collver (1977) could not enhance the solubility by employing sophisticated electron beam evaporation technique, thus limiting the solubility to a very low value between 3 to 4%.

Table 3.12 Structural parameters for $Si_{1-x}Ni_x$ through refinement of powder XRD data.

Parameter	Si	$Si_{0.97}Ni_{0.03}$	$Si_{0.94}Ni_{0.06}$	$Si_{0.91}Ni_{0.09}$	$Si_{0.88}Ni_{0.12}$
a (Å)	5.4371(3)	5.4236(2)	5.4116(2)	5.4032(3)	5.3911(3)
Volume (Å³)	160.73(9)	159.54(4)	158.48(7)	157.74(10)	156.68(8)
R_{obs} (%)	2.20	2.18	1.75	1.39	5.56
wR_{obs} (%)	2.35	1.94	2.16	1.10	4.89
R_p(%)	5.79	5.11	5.86	5.90	5.14
F_{000}	112	115	119	122	125
GoF	1.11	1.10	1.09	1.32	1.58
Grain size(nm)	55.59	34.98	40.14	37.40	34.76
Strain(x10^{-2})	0.1182	0.0427	0.0633	0.1534	0.2389
Particle size(μm)(SEM)	---	1.12	0.78	0.74	1.22

a-Cell dimension
R_{obs}-Reliability index for observed structure factors
wR_{obs}-Weighted reliability index for observed structure factors
R_p-Reliability index for profile
F_{000}-Number of electrons in the unit cell
GoF-Goodness of fit

Additional peaks were spotted at Bragg angles 44.5° and 52° for the prepared samples of $Si_{1-x}Ni_x$. This fact shows that, the additional phase of Ni exist along with Si:Ni system, even at very low dopant concentration *i.e.*, x = 0.03. As the concentration of the dopant is increased, the percentage of Ni cluster phase in the prepared material increases. Also, Ni was found to be included into the host Si matrix, which is evident from the comparison of intensity of (111) peak and the shift in the position of the peak. The intensity of the diffraction peaks increases, as the dopant concentration is increased in the host lattice, thus proving that, Ni occupies the substitutional position of the host lattice site of Si (Figure 3.19(b)). The higher atomic number of dopant atom Ni (Z = 28) than that of the host atom Si (Z = 14) causes an increase in the charge concentration at the host lattice site. Hence, as the dopant concentration is increased, there is an increase in the intensity of the diffraction peaks and the 2θ shift in the peaks. The electron beam evaporated Si (Collver, 1977) also shows the appearance of two phases, namely, pure Si and Ni metal element structure, in the form of micro clusters of pure Ni metal, along with Si:Ni system, where, Ni is a substitutional impurity. It was predicted that (Collver, 1977), the maximum percentage of dopant will be 14% up to which, Si preserves its structure, and hence, the doping percentage to go beyond 14% was deliberately avoided in our work.

(a)

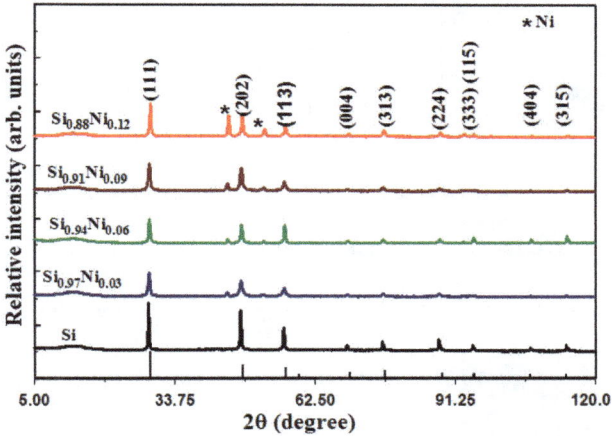

Figure 3.19 (a) Observed powder X-ray diffractogram of $Si_{1-x}Ni_x(x = 0.00, 0.03, 0.06, 0.09, 0.12)$.

(b)

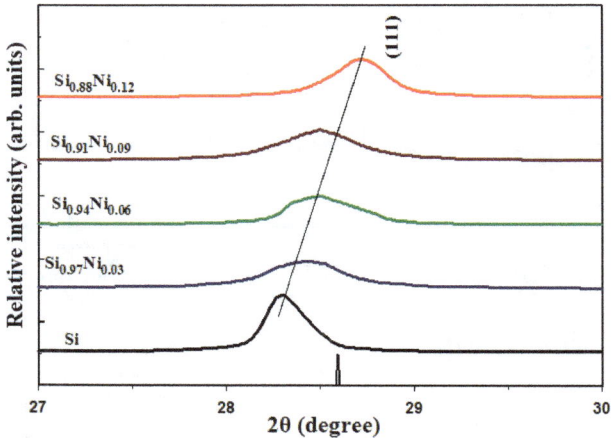

Figure 3.19 (b) Shift in the most prominent peak (111) in the observed powder X-ray diffractogram.

(a)

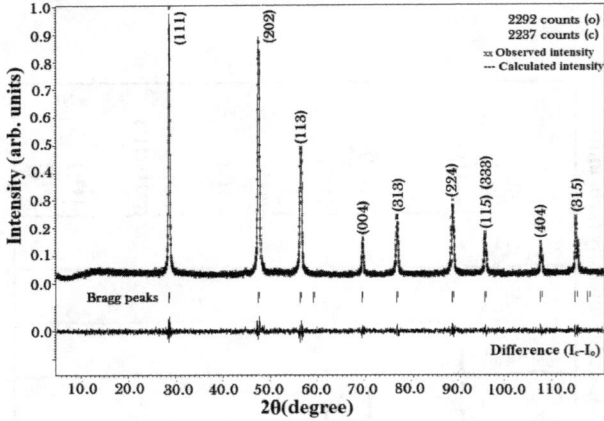

Figure 3.20 (a) Fitted powder XRD profile for Si.

(b)

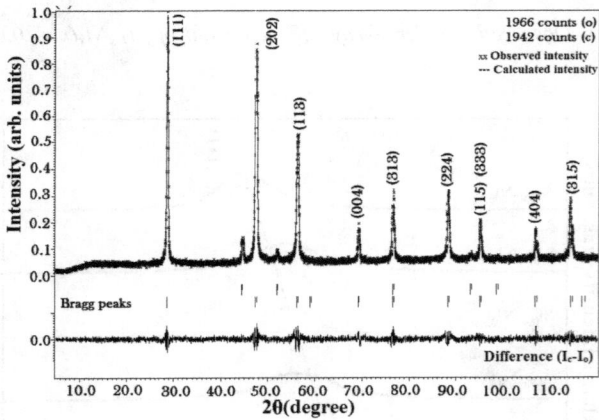

Figure 3.20 (b) Fitted powder XRD profile for $Si_{0.97}Ni_{0.03}$.

(c)

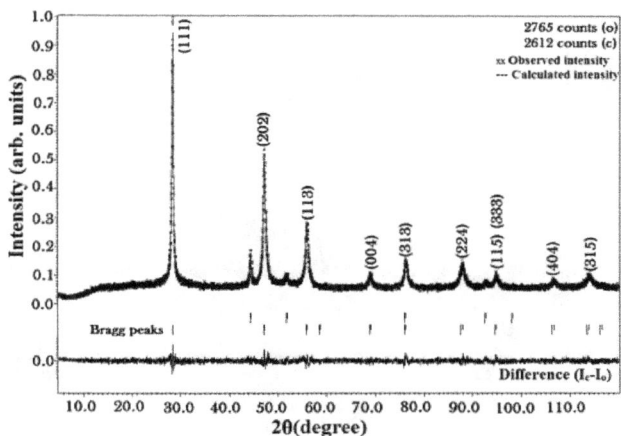

Figure 3.20 (c) Fitted powder XRD profile for $Si_{0.94}Ni_{0.06}$.

(d)

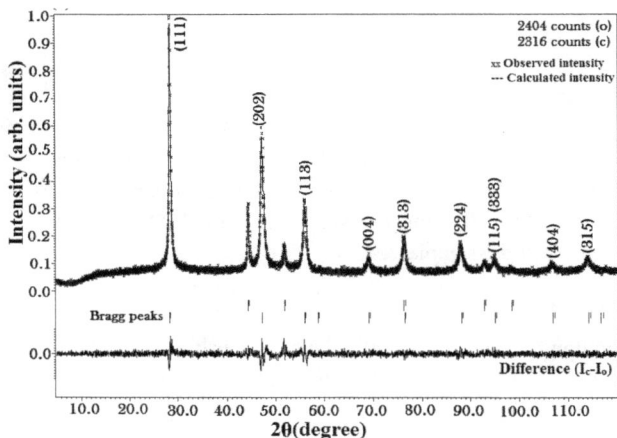

Figure 3.20 (d) Fitted powder XRD profile for $Si_{0.91}Ni_{0.09}$.

(e)

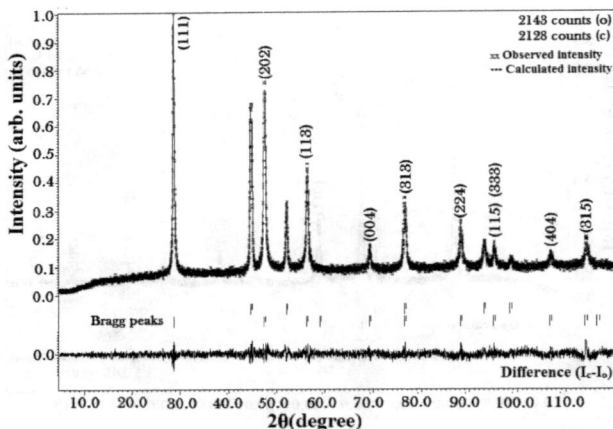

Figure 3.20 (e) Fitted powder XRD profile for $Si_{0.88}Ni_{0.12}$.

3.4 Conclusion

The melt grown samples of $Ge_{1-x}Mn_x$ (x = 0, 0.04, 0.06, 0.10), $Ge_{1-x}V_x$ (x = 0.03, 0.06, 0.09), $Ge_{1-x}Co_x$ (x = 0.03, 0.06, 0.09) and the ball milled samples of Si, $Si_{0.98}Mn_{0.02}$(100h), $Si_{0.98}Mn_{0.02}$(200h) and $Si_{1-x}Ni_x$ (x = 0, 0.03, 0.06, 0.09, 0.12) were characterized for their structural properties. The obtained powder X-ray diffraction data of the samples were refined using Rietveld (Rietveld, 1969) refinement, to obtain the structural parameters of the samples.

3.4.1 Melt grown $Ge_{1-x}Mn_x$

1. Incorporation of the dopant Mn in Ge host matrix is confirmed by the decrease in the total number of charges in the unit cell (F_{000}).

2. Compression and expansion of lattice is observed at x = 0.04 and x = 0.06respectively.

3. The reason for the compression and expansion of the lattice is found to be the hybridization of the orbitals of the dopant and the host atoms.

3.4.2 Melt grown $Ge_{1-x}V_x$

1. Intensity of the diffraction pattern is found to decrease with increase in the dopant concentration, which confirms the incorporation of vanadium atoms in the host lattice.

2. A decrease in the total number of charges in the unit cell (F_{000}) is observed with the increase in dopant concentration.

3. Distortion occurs at the host lattice sites, which is indicated by the shift in Bragg peaks towards the lower angles, as the dopant concentration is increased.

4. The values of the lattice constants and Debye-Waller factor (B_{iso}) increase with the dopant concentration, because of the presence of the static spin-density wave.

5. From the SEM micrographs, it is observed that, long range ordering is terminated and the crystalline nature is less for x = 0.03. Also, an improved crystalline nature for x = 0.06 and a non-uniform grain concentration for x = 0.09 is observed.

3.4.3 Melt grown $Ge_{1-x}Co_x$

1. For x = 0.03, the value of the lattice constant increases, when compared with that of undoped Ge.

2. For x = 0.06, the value of the cell constant decreases, whereas, for x = 0.09, the value of the lattice constant increases.

3. The total number of charges in the unit cell (F_{000}) decreases with increasing dopant concentration.

4. An additional phase of monoclinic GeCo is observed in the system and the intensity of the phase GeCo is found to increase with the dopant concentration, thus evidencing the inclusion of GeCo clusters in the prepared samples.

5. From the SEM micrographs of the sample with x = 0.03, it is observed that, the sample is with a good crystalline nature. A non-uniform grain concentration for x = 0.06 and a well pronounced crystallinity for x = 0.09 is observed.

3.4.4 Ball milled $Si_{1-x}Mn_x$

1. Inclusion of dopant Mn in the Si host lattice is confirmed by the increase in the cell dimension.

2. Decrease in the value of the Debye-Waller factor (B_{iso}) in the sample of $Si_{0.98}Mn_{0.02}$(100h), is attributed to the incorporation of heavier Mn atom in the host lattice of Si.

3. The heavy increase of the value of B_{iso} in the case of $Si_{0.98}Mn_{0.02}(200h)$ is due to the decrease in size of grain/domain, that accommodates the surface atoms and hence, the inner region has larger amplitude of vibration.

4. An additional phase of cubic SiMn, with the space group $P2_13$, is found to be present in the ball milled samples of $Si_{0.98}Mn_{0.02}$.

5. Decrease in the grain size with milling time, is evidenced from the SEM images of the samples.

3.4.5 Ball milled $Si_{1-x}Ni_x$

1. Inclusion of the dopant Ni in the host semiconductor matrix Si is confirmed by the shift in the Bragg peaks in the observed powder X-ray diffraction pattern.

2. A small fraction of the additional phase of pure Ni metal with FCC structure is found in the prepared Si:Ni system.

3. The decrease in the value of the cell dimension with the increase in the dopant concentration is attributed to the inclusion of the bigger Ni atom in the host Si matrix.

4. The increase in the agglomeration of the particles with the increase in the concentration of the dopant in the host lattice is observed. The homogeneity of the sample has been verified in the SEM measurements and hence a uniform experimental behavior is observed.

References

[1] Collver M.M., Solid State Communications, 23, 333 (1977). https://doi.org/10.1016/0038-1098(77)91340-0

[2] Haynes W.M., CRC Handbook of Chemistry and Physics, CRC Press/Taylor and Francis, Boca Raton, FL, 95th Edition, 2014.

[3] Joint Committee on Powder Diffraction Standards (JCPDS)PDF-2 Database PDF # 42-1487.

[4] Joint Committee on Powder Diffraction Standards (JCPDS)PDF-2 Database PDF # 65-1223

[5] Joint Committee on Powder Diffraction Standards (JCPDS)PDF-2 Database PDF # 04-0850.

[6] Mocking T.F., Hlawacek G., Zandvliet H.J.W., Surface Science, 606, 924(2012). https://doi.org/10.1016/j.susc.2012.02.007

[7]   Petříček V., Dušek M., Palatinus L., Zeitschrift für Kristallographie-Crystalline
       Materials, 229 (5), 345(2014). https://doi.org/10.1515/zkri-2014-1737

[8]   Rietveld H.M., Journal of Applied Crystallography, 2, 65(1969).
       https://doi.org/10.1107/S0021889869006558

[9]   Saravanan R., FORTRAN program 'GRAIN' www.saraxraygroup.net.

[10]   Slater J.C., J. Chem. Phys.,41, 3199 (1964). https://doi.org/10.1063/1.1725697

[11]   Weng H., Dong J., Phys. Rev. B, 71, 035201(2005).
       https://doi.org/10.1103/PhysRevB.71.035201

[12]   Wyckoff R.W.G., Crystal Structure, London (Inter-Science publishers), Vol. I,
       1963.

Dilute Magnetic Semiconducting (DMS) Materials, R. Saravanan   Materials Research Forum LLC
Materials Research Foundations **35** (2018)   doi: http://dx.doi.org/10.21741/9781945291777

Chapter 4

Theoretical Electronic Charge Density Distribution

Abstract

Chapter 4 gives a detailed analysis of the theoretical charge density distribution of the proposed materials. This theoretical estimation of the charge density of the materials is useful to predict the properties of the DMS materials before moving on to the experimental understanding. The charge density of the Ge and Si based DMS materials with the dopants of manganese (Mn), cobalt (Co) and vanadium (V) is estimated theoretically and analyzed using the charge density maps. The results are presented in this chapter.

Keywords

Theoretical, Charge Density, Si and Ge Based, Transition Metals

Contents

4.1   Introduction on theoretical electronic charge density distribution

In order to achieve successful fabrication of materials suitable for specific application purposes, there is a need to understand the properties of the materials. Therefore, a prior theoretical estimation on charge related properties would be very helpful. By estimating

Dilute Magnetic Semiconducting (DMS) Materials, R. Saravanan   Materials Research Forum LLC
Materials Research Foundations **35** (2018)   doi: http://dx.doi.org/10.21741/9781945291777

the charge density of a particular material, the bonding nature and hence the properties of the material can be very well comprehended. In a crystal, the interaction of the charges can be determined using X-rays through which, the structure of the crystal can be understood. The determination of the structure of the crystal yields information on lattice arrangement of atoms, lengths and angles of the bonding, type of bonding, the effect of impurities and the contribution of doping to the crystal structure and cell parameters. The physical, electrical and thermal properties of the materials cannot be understood using the structural information alone. Hence, the analysis of the electronic charge density distribution becomes inevitable. The quantitative measurement of the charges involved in the bonding and the types of bonding between the atoms can be done using the charge density analysis.

Here, an attempt has been made to understand the ideal charge density distribution of silicon and germanium based diluted magnetic semiconductors using a theoretical route. For the theoretical estimation of the electronic charge density, silicon (Si) and germanium (Ge) are chosen as host materials. The transition metals vanadium (V), manganese (Mn) and cobalt (Co) are chosen as the dopants. The detailed theoretical charge density analysis of the chosen systems were done and presented here.

4.2 Theoretical charge density estimation of DMS materials $Si_{1-x}M_x$ and $Ge_{1-x}M_x$ (M = V, Mn, Co)

The theoretical charge density of the DMS materials $Si_{1-x}M_x$ and $Ge_{1-x}M_x$ is estimated as follows.

(a) Calculation of structure factors

The structure factor of a material is given by Eqn. (2.6) and Eqn. (2.7) as:

$$F_{hkl} = \sum_{j=1}^{m} N_j f_j \, exp\big[2\pi i\big(hx_j + ky_j + lz_j\big)\big] \tag{2.6}$$

where j refers to all the atoms in the unit cell and f_j is calculated from

$$|f|^2 = \left(f_0 \, exp\left[\frac{-Bsin^2\theta}{\lambda^2}\right] + \Delta f'\right)^2 + (\Delta f'')^2 \tag{2.7}$$

where f_0 can be determined by the method developed by Doyle and Turner (1968) who used linear combinations of four Gaussians:

$$f_0(s) = \sum_{i=1}^{4} a_i \exp(-b_i s^2) + c \tag{4.1}$$

which have been found to give close fits to the tabulated values up to $s_{max} = 2.0\ \text{Å}^{-1}$. Eqn. (4.1) and the parameters a_i, b_i and c are compiled in Table 6.1.1.4 of International Tables for Crystallography (1992), Vol. C which is implemented in the program sfac332 (Saravanan).

The steps followed in the code are given below:

1. The atoms involved in the crystal system are identified. In our case, for $A_{1-x}B_x$, A = Si, Ge and B = Mn, V, Co.

2. The general positions of the atoms involved in the system are considered as (x_j, y_j, z_j). In our case, the atoms A and B assume the same positions, while the percentage of the dopant concentration is noted according to the requirement.

3. The list of allowed *(hkl)* values is prepared for a particular system. The corresponding values of $\sin\theta/\lambda$ are estimated. The Debye-Waller factor B for the atoms in the system is selected according to the experimental findings. Then, f_0 is calculated using Eqn. (4.1).

4. For Eqn. (4.1), s is selected from 0 to 2.0Å^{-1} in steps of 0.05. The values of a_i, b_i and c along with $\Delta f'$ and $\Delta f''$ are used from the International Tables (1992) as analytical scattering factor functions and the values for the host and the dopant elements are tabulated in table 4.1. These values are incorporated into Eqn. (4.1) and $f_0(s)$ is found for each value of s where $s = \sin\theta/\lambda$.

5. Then, $|f|^2$ is estimated from Eqn. (2.7) and the square root of it is fit into Eqn. (2.6) for the calculation of F_{hkl}. In case of a doped system, F_j in Eqn. (2.6) is specified as $f_j(A_{1-x}) + f_j(B_x)$.

6. Similarly, for all the *(hkl)* values, the structure factors F_{hkl} are determined.

(b) Estimation of the theoretical charge density

The charge density of the DMS materials is estimated as follows:

1. For the estimation of the charge density distribution, the values $|F_{hkl}|$ are calculated and the corresponding real and imaginary parts are calculated separately as A(**H**) and B(**H**) respectively.

2. Now, the unit cell is divided into several pixels and the position of each pixel is assigned as r.

3.   The charge density $\rho(r)$ is calculated according to Eqn. (2.12) for all pixel points in the unit cell.

$$\rho(r) = \frac{1}{V}\sum[A(H) + iB(H)][\cos(2\pi H \cdot r) - i\sin(2\pi H \cdot r)] \qquad (2.12)$$

Table 4.1 Analytical scattering factor functions*.

Atoms	Si	Ge	Mn	V	Co
a_1	5.275329	16.540614	11.709542	`10.473575	12.914510
b_1	2.631338	2.866618	5.597120	7.081940	4.507138
a_2	3.191038	1.567900	1.733414	1.547881	2.481908
b_2	33.730728	0.012198	0.017800	0.026040	0.009126
a_3	1.511514	3.727829	2.673141	1.986381	3.466894
b_3	0.081119	13.432163	21.788419	31.909672	16.438130
a_4	1.356849	3.345098	2.023368	1.865616	2.106351
b_4	86.288640	58.866046	89.517915	108.022844	76.987317
a_5	2.519114	6.785079	7.003180	7.056250	6.960892
b_5	1.170087	0.210974	0.383054	0.474882	0.314418
c	0.145073	0.018726	-0.147293	0.067744	-0.936572
$\Delta f'$	0.2541	-1.0885	-0.5299	0.0687	-2.3653
$\Delta f''$	0.3302	0.8855	2.8052	2.1097	3.6143

* - From International Tables (1992)

4.2.1 Theoretical charge density estimation of DMS materials $Si_{1-x}M_x$ (M = V, Mn, Co)

The structure factors of the systems of $Si_{1-x}M_x$ (M = V, Mn, Co; x = 0.02, 0.04, 0.06, 0.08, 0.10) were calculated using the FORTRAN program 'sfac332' (Saravanan). Silicon (Si) was considered as the host material having the diamond structure and the space group $Fd\bar{3}m$. The cell parameters were assumed to be a = b = c = 5.4309 Å and the angles were α = β = γ = 90° (Wyckoff, 1963). The structure factors were calculated for the possible 72 Bragg reflections of Si system and used for the estimation of the electronic charge density distribution using the maximum entropy method (Collins, 1982). The charge densities of the systems were estimated using the software package

Dilute Magnetic Semiconducting (DMS) Materials, R. Saravanan   Materials Research Forum LLC
Materials Research Foundations **35** (2018)   doi: http://dx.doi.org/10.21741/9781945291777

PRIMA (PRactice of Interactive MEM Analysis) (Izumi et al., 2002) and the estimated charge density distributions were mapped using the visualization software VESTA (Visualization for Electronic and STructural Analysis) (Momma et al., 2011).

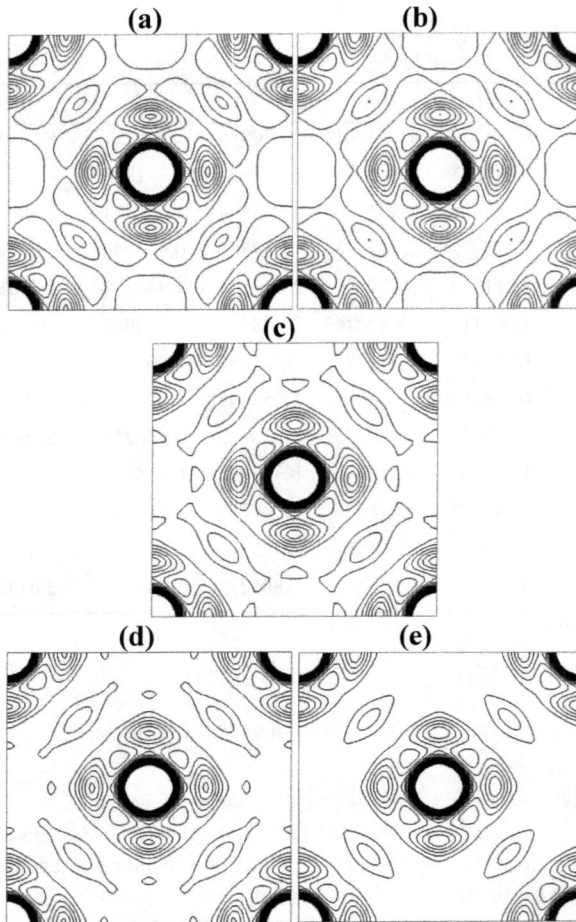

Figure 4.1 Theoretical two dimensional electron density distributionon (100) plane for $Si_{1-x}V_x$(a) x = 0.02 (b) x = 0.04 (c) x = 0.06 (d) x = 0.08 (e) x = 0.10 (Contour range: 0.2-2 e/$\mathring{A}^3$; Contour interval: 0.05e/$\mathring{A}^3$).

Dilute Magnetic Semiconducting (DMS) Materials, R. Saravanan   Materials Research Forum LLC
Materials Research Foundations **35** (2018)   doi: http://dx.doi.org/10.21741/9781945291777

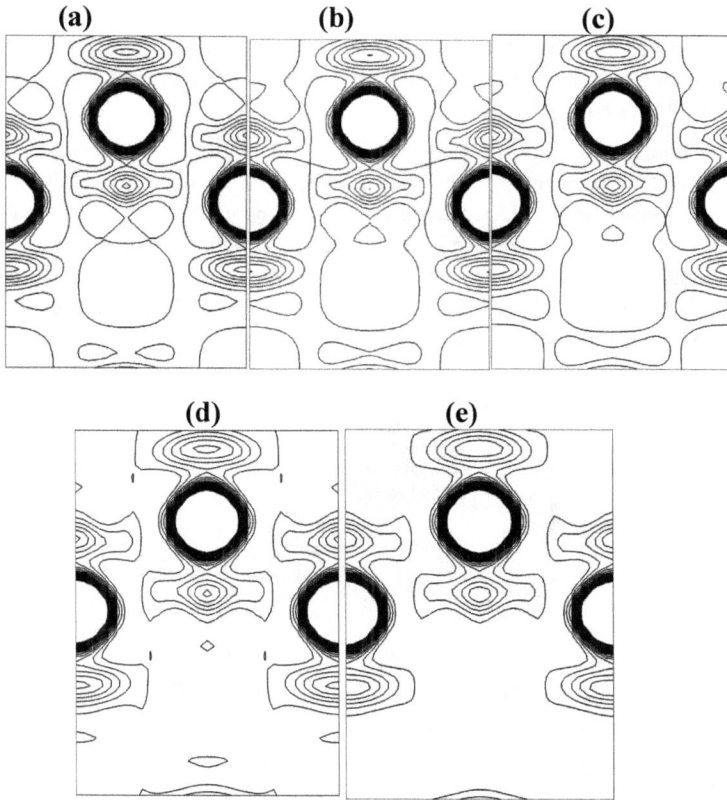

Figure 4.2 Theoretical two dimensional electron density distributionon (110) plane for $Si_{1-x}V_x$(a) x = 0.02 (b) x = 0.04 (c) x = 0.06 (d) x = 0.08 (e) x = 0.10 (Contour range: 0.2-2 e/$Å^3$; Contour interval: 0.05 e/$Å^3$).

The two dimensional electronic charge density distributions for the systems of $Si_{1-x}V_x$, $Si_{1-x}Mn_x$ and $Si_{1-x}Co_x$ were mapped on the planes (100) and (110) with the contour level between 0.2 e/$Å^3$ and 2.0 e/$Å^3$ with the interval of 0.05 e/$Å^3$.

The two dimensional charge density mapping of the system $Si_{1-x}V_x$ (x = 0.02, 0.04, 0.06, 0.08, 0.10) on the planes (100) and (110) are shown in figures 4.1(a) to (e) and figures 4.2 (a) to (e) respectively. From the two dimensional maps, it is evident that the contour lines become less dense as the concentration of the dopant increases. The electron density between the corner atoms and the face centered atoms become weaker as the dopant concentration increases. This can also be seen clearly on the plane (110) (Figures4.2 (a) to (e)).

The one dimensional charge density profiles of $Si_{1-x}V_x$ for the five compositions of x = 0.02, 0.04, 0.06, 0.08 and 0.10 are shown in figure 4.3. It can be clearly seen that, the mid bond density of $Si_{1-x}V_x$ decreases when the value of the concentration of the dopant increases. This phenomenon can be attributed to the higher atomic number of the dopant atom vanadium (Z = 23) than that of the host atom silicon (Z = 14). Because of the higher atomic number of the dopant, as the doping concentration increases, the attraction to the core increases thereby decreasing the mid bond density.

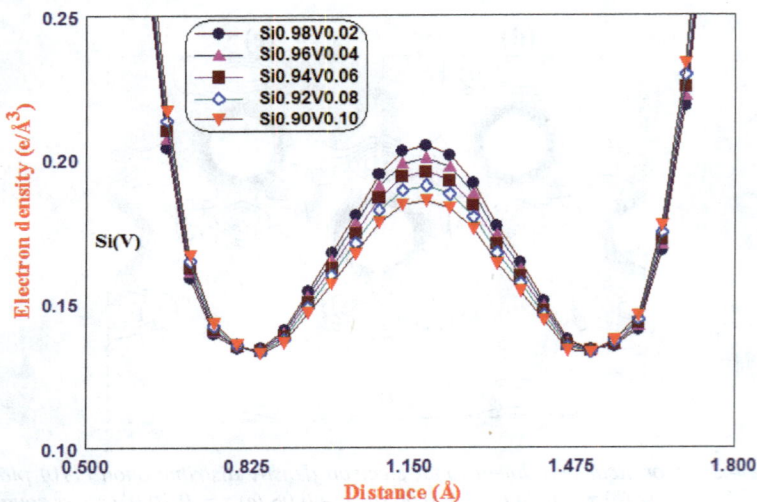

Figure 4.3 Theoretical one dimensional charge density profile of $Si_{1-x}V_x$ (x = 0.02, 0.04, 0.06, 0.08, 0.10) along the [111] direction.

The two dimensional electron density maps of $Si_{1-x}Mn_x$ (x = 0.02, 0.04, 0.06, 0.08, 0.10) mapped on the (100) plane are shown in figures 4.4 (a) to (e). The two dimensional charge density mapping on the plane (110) is shown in figures 4.5 (a) to (e). It is evident from the two dimensional electron density maps that, as the concentration of the dopant (Mn) increases, the valence charges are attracted towards the core. The electron density in the system decreases as the concentration of the dopant increases which is evident from the sparingly spaced contour lines.

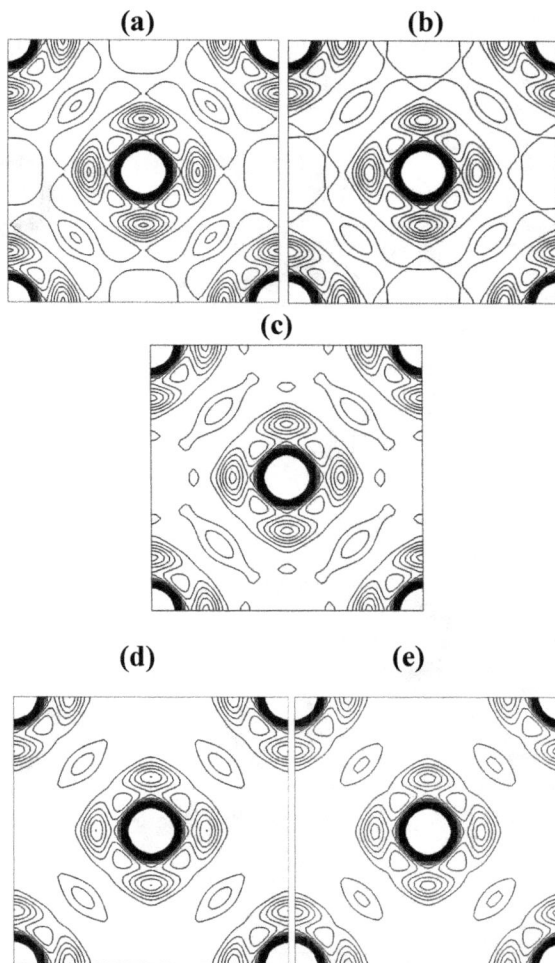

Figure 4.4 Theoretical two dimensional electron density distributionon (100) plane for
$Si_{1-x}Mn_x(a)$ *x = 0.02 (b) x = 0.04 (c) x = 0.06 (d) x = 0.08 (e) x = 0.10 (Contour range:*
0.2-2 e/$Å^3$; Contour interval: 0.05 e/$Å^3$).

The one dimensional electron density profiles are mapped for the system of $Si_{1-x}Mn_x$ (x = 0.02, 0.04, 0.06, 0.08, 0.10) along the bonding direction of [111] and compared in figure 4.6.It is evident that, the density of the charge distribution decreases as the dopant concentration increases.

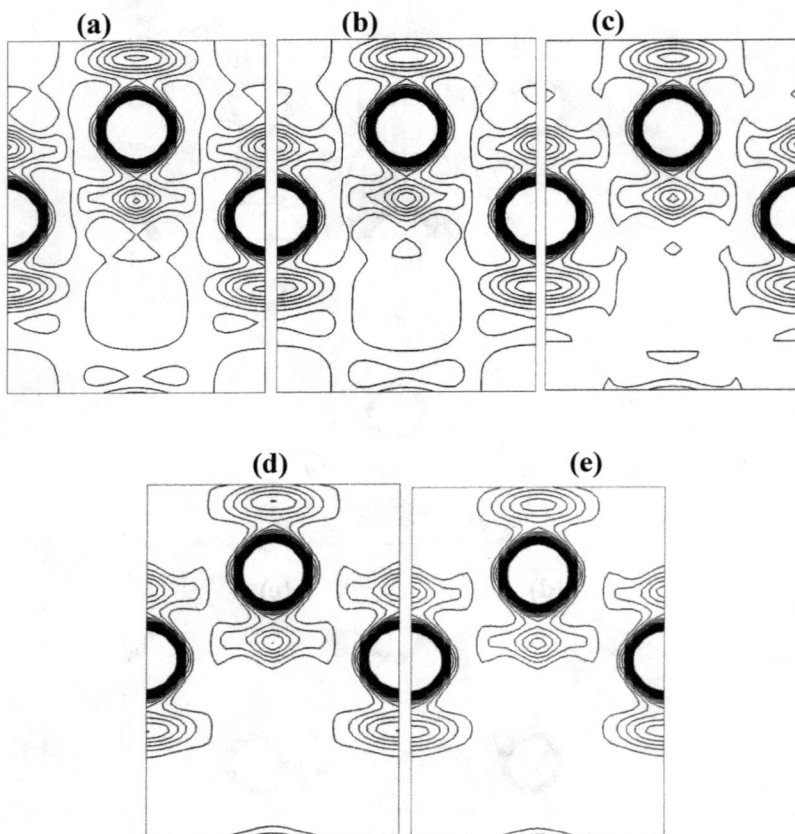

Figure 4.5 Theoretical two dimensional electron density distributionon (110) plane for $Si_{1-x}Mn_x$ (a) x = 0.02 (b) x = 0.04 (c) x = 0.06 (d) x = 0.08 (e) x = 0.10 (Contour range: 0.2-2 e/Å³; Contour interval: 0.05 e/Å³).

The two dimensional electron density maps of $Si_{1-x}Co_x$ (x = 0.02, 0.04, 0.06, 0.08, 0.10) mapped on the (100) plane are shown in figures 4.7 (a) to (e). The two dimensional charge density mapping on the plane (110) is shown in figures 4.8 (a) to (e). The mapped two dimensional charge densities for the samples show that the electron density decreases as the concentration of the dopant in the host lattice increases.

The one dimensional electron density profiles are mapped for the system of $Si_{1-x}Co_x$ (x = 0.02, 0.04, 0.06, 0.08, 0.10) along the bonding direction of [111] and compared in figure 4.9. The mid bond density decreases as the concentration of the dopant is increased in the host silicon lattice since the atomic number of the dopant Co (Z = 27) is higher than that of the host Si (Z = 14).

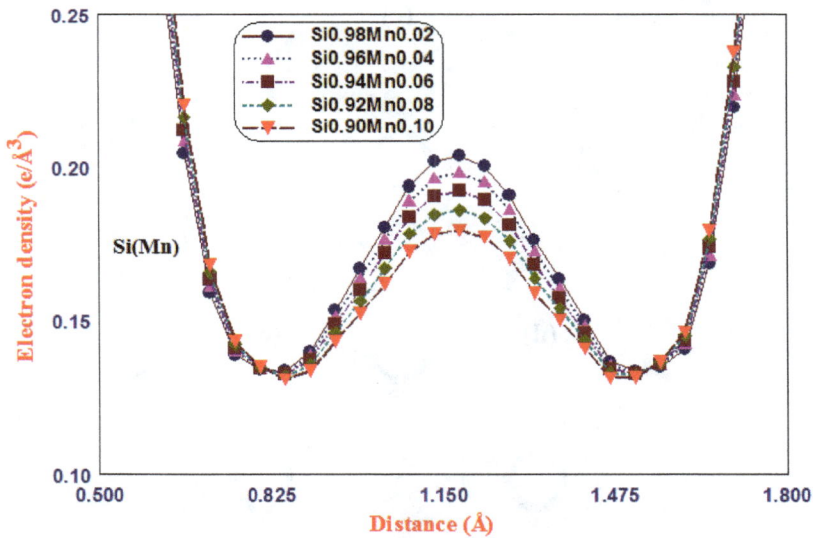

Figure 4.6 Theoretical one dimensional charge density profile of $Si_{1-x}Mn_x$ (x = 0.02, 0.04, 0.06, 0.08, 0.10) along the [111] direction.

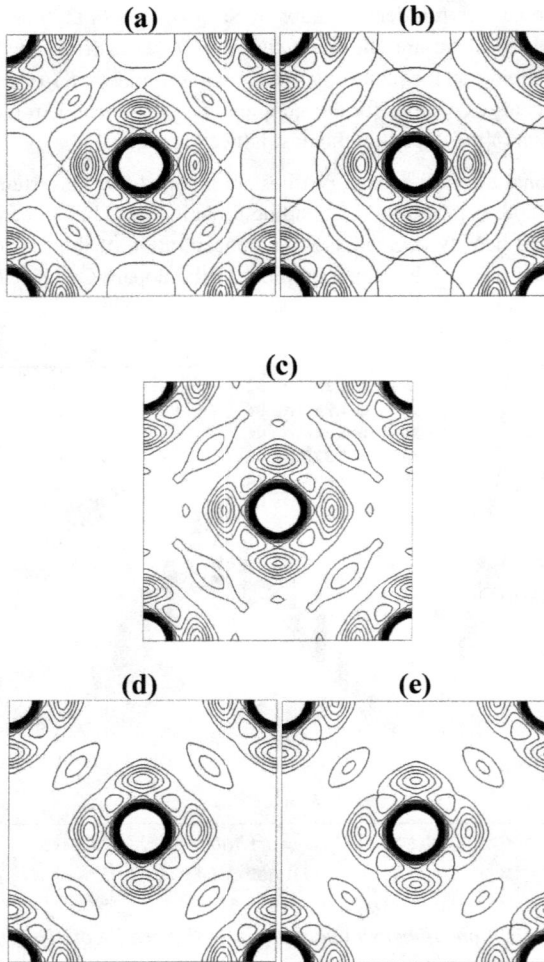

Figure 4.7 Theoretical two dimensional electron density distributionon (100) plane for Si$_{1-x}$Co$_x$(a) x = 0.02 (b) x = 0.04 (c) x = 0.06 (d) x = 0.08 (e) x = 0.10 (Contour range: 0.2-2 e/Å^3; Contour interval: 0.05 e/Å^3).

Dilute Magnetic Semiconducting (DMS) Materials, R. Saravanan
Materials Research Forum LLC
Materials Research Foundations **35** (2018)
doi: http://dx.doi.org/10.21741/9781945291777

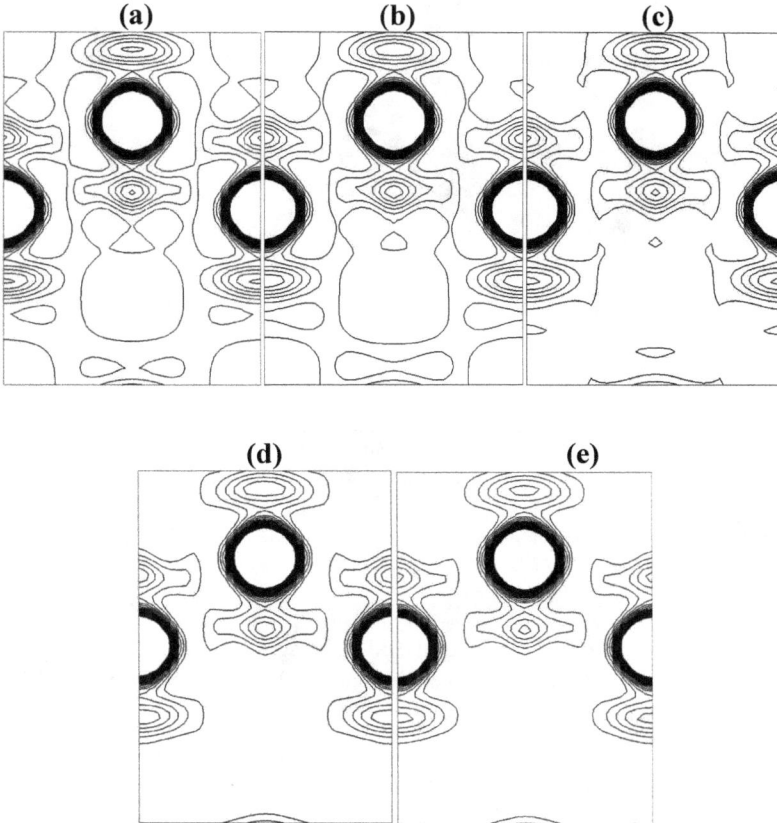

Figure 4.8 Theoretical two dimensional electron density distributionon (110) plane for $Si_{1-x}Co_x$*(a) x = 0.02 (b) x = 0.04 (c) x = 0.06 (d) x = 0.08 (e) x = 0.10 (Contour range:* 0.2-2 $e/Å^3$*; Contour interval:* 0.05 $e/Å^3$*).*

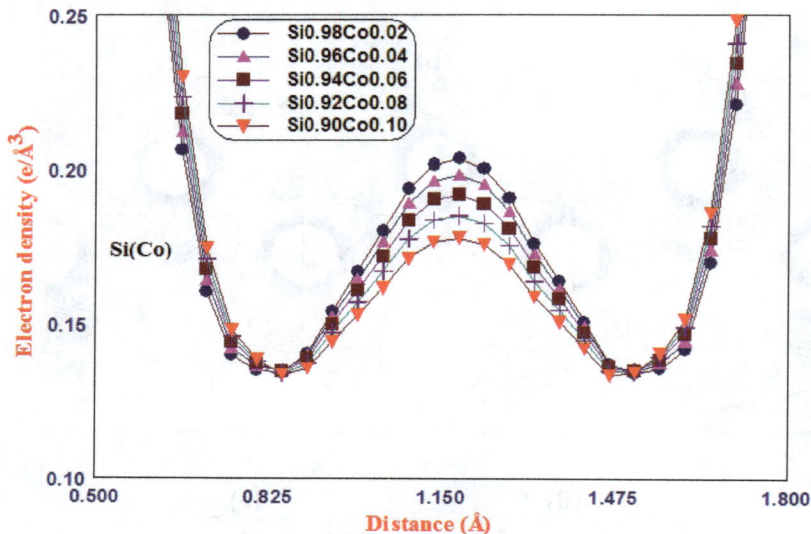

Figure 4.9 Theoretical one dimensional charge density profile of $Si_{1-x}Co_x$ (x = 0.02, 0.04, 0.06, 0.08, 0.10) along the [111] direction.

The mid bond densities of the system of the diluted magnetic semiconductors $Si_{1-x}V_x$, $Si_{1-x}Mn_x$ and $Si_{1-x}Co_x$ for all the five compositions of x = 0.02, 0.04, 0.06, 0.08 and 0.10 are compared and presented in table 4.2.

It is clear that (table 4.2), as the concentration of the transition metal atom (dopant atom of V, Mn or Co) increases, the mid bond densities of the diluted magnetic semiconductor along the bonding direction of [111] decreases. The atomic numbers of the transition metals vanadium (V), manganese (Mn) and cobalt (Co) are 23, 25 and 27 respectively. When the atomic number of the dopant is less, the incorporation of the dopant atoms is less in the substitutional sites. This results in less attraction of electronic charges towards the core. The bonding in the host system (Si) is covalent in which electron sharing takes place. When the atomic weight of the core is less, the attraction towards the core is less. Hence, more number of electrons can be shared which leads to the higher value of the mid bond density. But, when the atomic number of the dopant is high (27 for Co), the supply to the nucleus in the substitutional sites is more and hence the attraction towards the core is more. This leads to the decrease in the mid bond density. Hence, the mid bond electron charge density for the dopant vanadium is more when compared to the other two dopant atoms Mn and Co.

Table 4.2 Comparison of the theoretical mid bond densities along the bonding direction [111] for the samples of $Si_{1-x}M_x$ (M = V, Mn, Co).

Composition (x)	Bond length (Å)	Electron density (e/Å³)		
		$Si_{1-x}V_x$	$Si_{1-x}Mn_x$	$Si_{1-x}Co_x$
0.02	1.1817	0.2044	0.2035	0.2035
0.04	1.1817	0.2001	0.1980	0.1980
0.06	1.1817	0.1954	0.1921	0.1918
0.08	1.1817	0.1906	0.1858	0.1850
0.10	1.1817	0.1855	0.1791	0.1778

4.2.2 Theoretical charge density estimation of DMS material $Ge_{1-x}M_x$ (M = V, Mn, Co)

The structure factors of the systems of $Ge_{1-x}M_x$ (M = V, Mn, Co; x = 0.02, 0.04, 0.06, 0.08, 0.10) were calculated using the FORTRAN program 'sfac332' (Saravanan). Germanium (Ge) having the diamond structure and the space group $Fd\bar{3}m$ was considered as the host material. The cell parameters were assumed to be a = b = c = 5.667 Å and the angles were α = β = γ = 90° (Wyckoff, 1963). The structure factors were calculated for the possible 91 Bragg reflections of Ge system and used for the estimation of the electronic charge density distribution using the maximum entropy method (Collins, 1982). The charge densities of the systems were estimated using the software package PRIMA (PRactice of Interactive MEM Analysis) (Izumi et al., 2002) and the estimated charge density distributions were mapped using the visualization software VESTA (Visualization for Electronic and STructural Analysis) (Momma et al., 2011).

The two dimensional electronic charge density distributions for the systems of $Ge_{1-x}V_x$, $Ge_{1-x}Mn_x$ and $Ge_{1-x}Co_x$ were mapped on the planes (100) and (110) with the contour level between 0.2 e/Å³ and 2.0 e/Å³ with the interval of 0.1 e/Å³.

The two dimensional electron densities for the system $Ge_{1-x}V_x$ (x = 0.02, 0.04, 0.06, 0.08, 0.10) mapped on the planes (100) and (110) are shown in figures 4.10 (a) to (e) and figures 4.11 (a) to (e) respectively. It is evident that, as the concentration of the dopant

(V) in the host lattice increases, the mid bond density between the atoms increases. This increase in the mid bond density is due to the higher atomic number of the host Ge (Z = 32) than the dopant V (Z = 23).

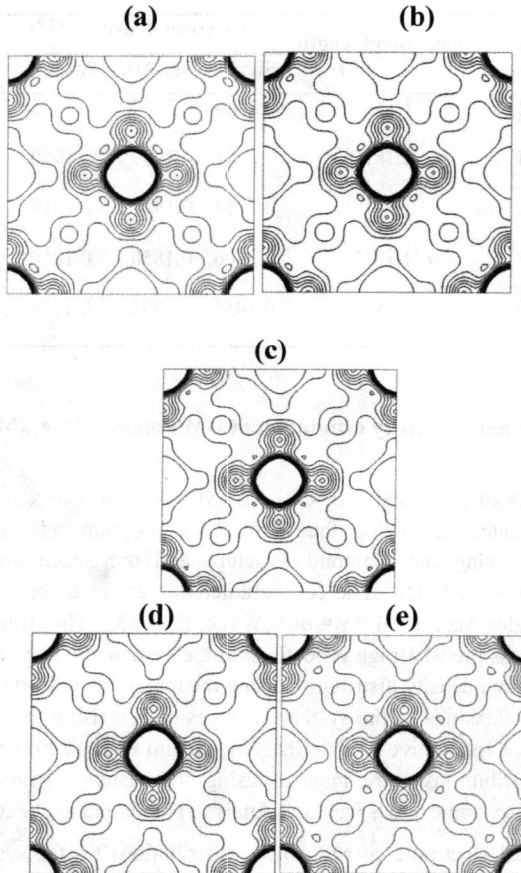

(a)   **(b)**

(c)

(d)   **(e)**

Figure 4.10 Theoretical two dimensional electron density distributionon (100) plane for Ge$_{1-x}$V$_x$(a) x = 0.02 (b) x = 0.04 (c) x = 0.06 (d) x = 0.08 (e) x = 0.10 (Contour range: 0.2-2 e/Å^3; Contour interval: 0.05 e/Å^3).

Dilute Magnetic Semiconducting (DMS) Materials, R. Saravanan   Materials Research Forum LLC
Materials Research Foundations **35** (2018)   doi: http://dx.doi.org/10.21741/9781945291777

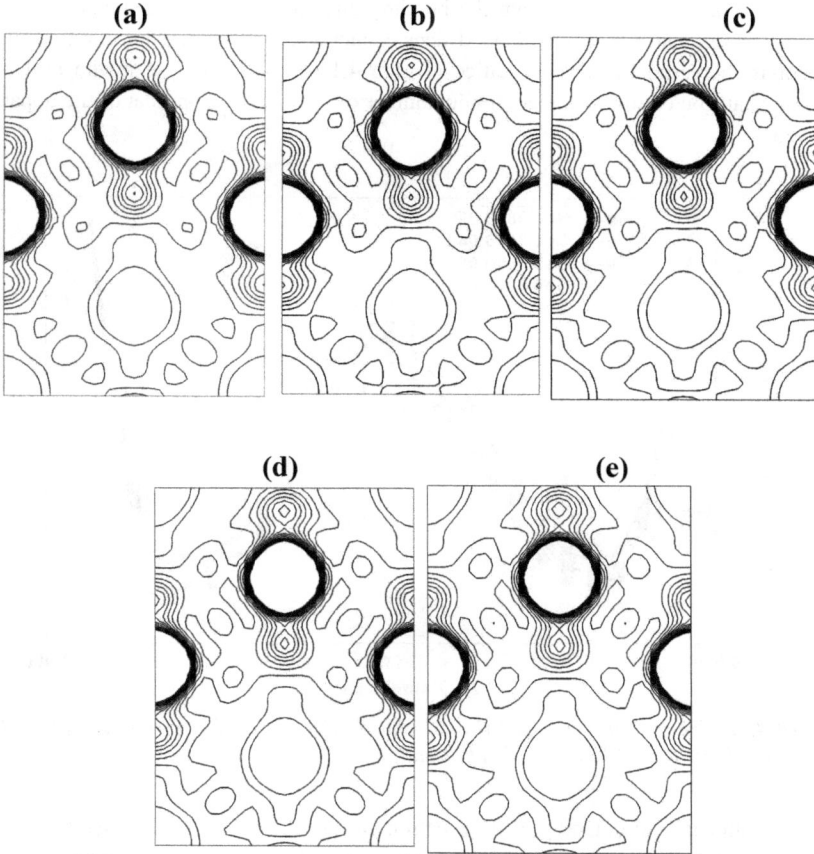

Figure 4.11 Theoretical two dimensional electron density distributionon (110) plane for
$Ge_{1-x}V_x(a)$ *x = 0.02 (b) x = 0.04 (c) x = 0.06 (d) x = 0.08 (e) x = 0.10 (Contour range:*
0.2-2 $e/Å^3$; Contour interval: 0.05 $e/Å^3$).

The one dimensional electron density profiles are mapped for the system of $Ge_{1-x}V_x$ (x = 0.02, 0.04, 0.06, 0.08, 0.10) along the bonding direction [111] and compared in figure 4.12. It is evident that, the mid bond density increases when the concentration of the dopant is increased in the host lattice (Figure 4.12). This increase in the mid bond density is attributed to the higher atomic number of the host Ge than that of the dopant V (Z = 23).

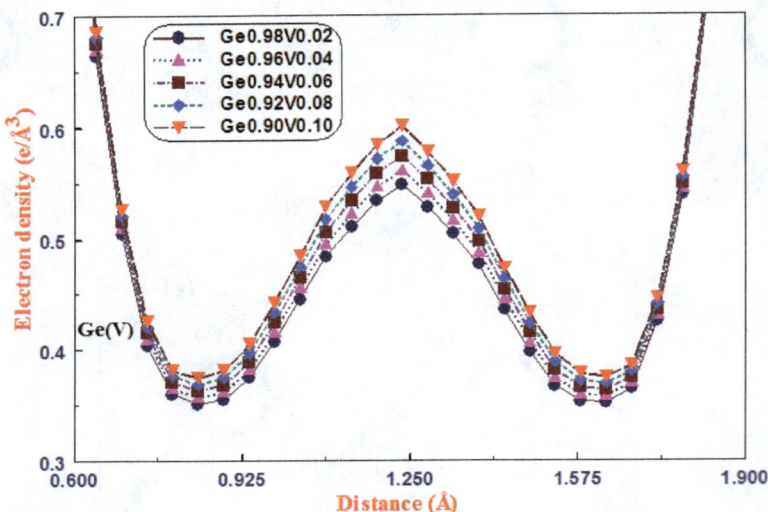

Figure 4.12 Theoretical one dimensional charge density profile of $Ge_{1-x}V_x$ (x = 0.02, 0.04, 0.06, 0.08, 0.10) along the [111] direction.

The two dimensional mapping of the electron density distribution of the system of $Ge_{1-x}Mn_x$ (x = 0.02, 0.04, 0.06, 0.08, 0.10) on the planes (100) and (110) are shown in figures 4.13 (a) to (e) and figures 4.14 (a) to (e) respectively. It is evident that, the increase in the dopant (Mn) concentration in the host lattice causes an increase in the electron density between the atoms in the lattice since the atomic number of the dopant Mn (Z = 25) is lower than the host Ge (Z = 32).

The one dimensional electronic charge density profiles of the system of $Ge_{1-x}Mn_x$ (x = 0.02, 0.04, 0.06, 0.08, 0.10) along the bonding direction [111] are mapped and shown in figure 4.15. It is clear that, when the concentration of the dopant is high, the density of

the charges shared for the bonding is high. Also, when the concentration is low, less
charge is shared between the atoms in the lattice (Figure 4.15).

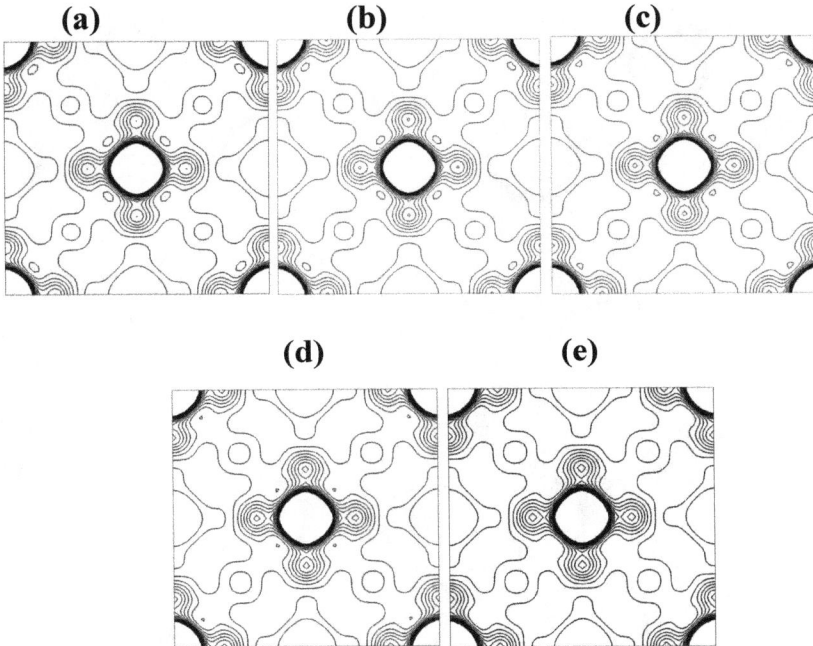

(a)   **(b)**   **(c)**

(d)   **(e)**

Figure 4.13 Theoretical two dimensional electron density distribution on (100) plane for
$Ge_{1-x}Mn_x$(a) x = 0.02 (b) x = 0.04 (c) x = 0.06 (d) x = 0.08 (e) x = 0.10 (Contour range:
0.2-2 $e/\mathring{A}^3$; Contour interval: 0.05 $e/\mathring{A}^3$).

Dilute Magnetic Semiconducting (DMS) Materials, R. Saravanan   Materials Research Forum LLC
Materials Research Foundations **35** (2018)   doi: http://dx.doi.org/10.21741/9781945291777

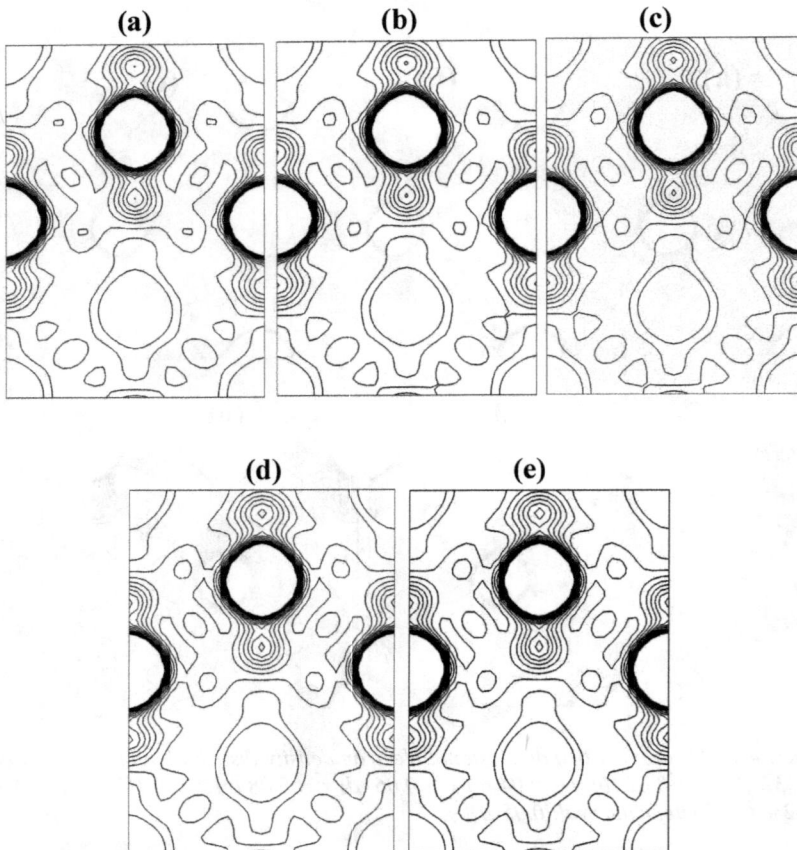

Figure 4.14 Theoretical two dimensional electron density distributionon (110) plane for $Ge_{1-x}Mn_x$*(a) x = 0.02 (b) x = 0.04 (c) x = 0.06 (d) x = 0.08 (e) x = 0.10 (Contour range: 0.2-2 e/Å^3; Contour interval: 0.05 e/Å^3).*

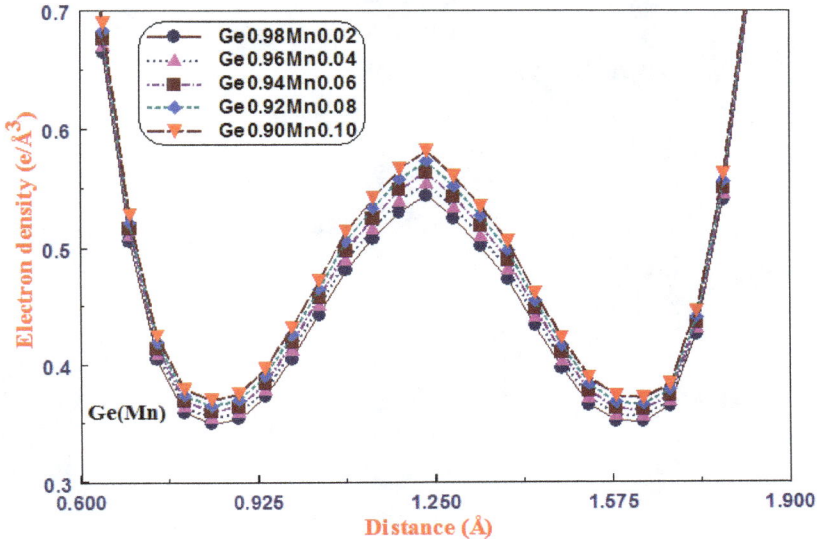

Figure 4.15 Theoretical one dimensional charge density profile of $Ge_{1-x}Mn_x$ (x = 0.02, 0.04, 0.06, 0.08, 0.10) along the [111] direction.

The theoretical estimation of the charge densities was also carried out using cobalt (Co) as the dopant in the host lattice of germanium (Ge). The mapping of the charge densities on the planes (100) and (110) is shown in figures 4.16 (a) to (e) and figures 4.17 (a) to (e) respectively. An increase in the charge density in the mid bond region between the corner and the face centered atoms can be clearly seen when the concentration of the dopant is increased.

The one dimensional electronic charge density profiles for the systems of $Ge_{1-x}Co_x$ (x = 0.02, 0.04, 0.06, 0.08, 0.10) are mapped along the bonding direction [111] and are compared in figure 4.18. It is substantiated in the one dimensional electron density profiles that, the mid bond density between the corner and the face centered atoms increases as the concentration of the dopant increases.

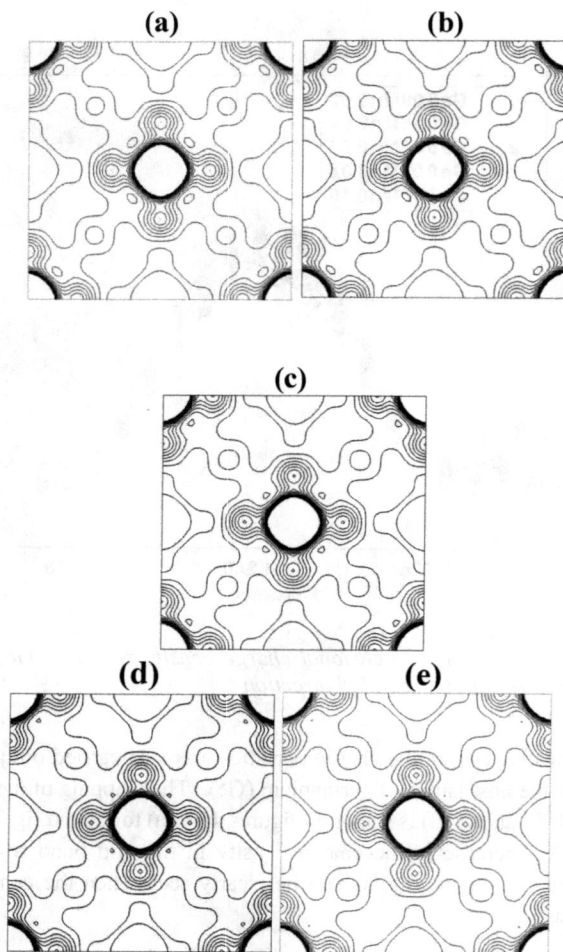

Figure 4.16 Theoretical two dimensional electron density distribution on (100) plane for $Ge_{1-x}Co_x$ *(a) x = 0.02 (b) x = 0.04 (c) x = 0.06 (d) x = 0.08 (e) x = 0.10 (Contour range: 0.2-2 e/Å³; Contour interval: 0.05 e/Å³).*

Dilute Magnetic Semiconducting (DMS) Materials, R. Saravanan   Materials Research Forum LLC
Materials Research Foundations **35** (2018)   doi: http://dx.doi.org/10.21741/9781945291777

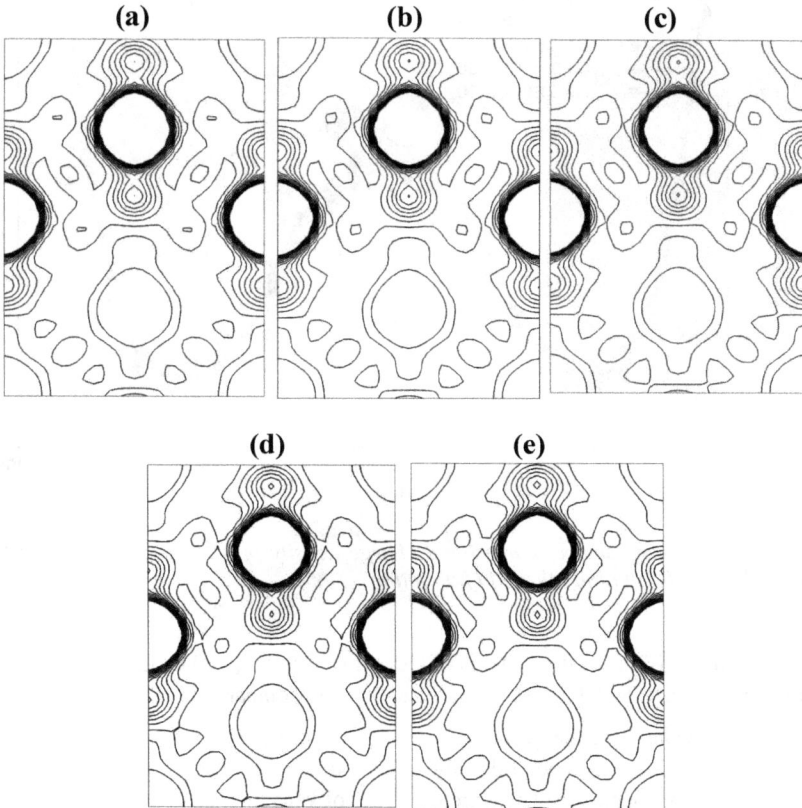

Figure 4.17 Theoretical two dimensional electron density distribution on (110) plane for Ge$_{1-x}$Co$_x$ (a) x = 0.02 (b) x = 0.04 (c) x = 0.06 (d) x = 0.08 (e) x = 0.10 (Contour range: 0.2-2 e/Å^3; Contour interval: 0.05 e/Å^3).

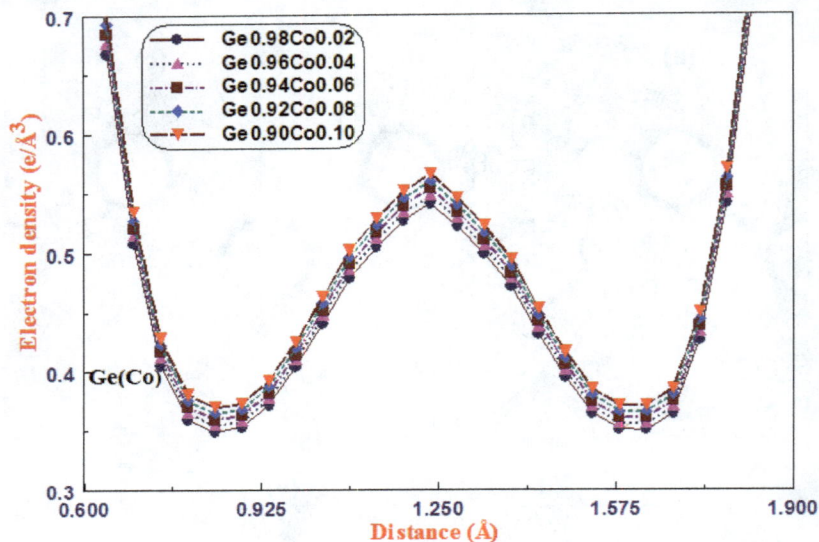

Figure 4.18 Theoretical one dimensional charge density profile of Ge$_{1-x}$Co$_x$ (x = 0.02, 0.04, 0.06, 0.08, 0.10) along the [111] direction.

The mid bond densities of the systems of Ge$_{1-x}$Mn$_x$, Ge$_{1-x}$Co$_x$ and Ge$_{1-x}$V$_x$ for all the five compositions of x = 0.02, 0.04, 0.06, 0.08 and 0.10 are compared in table 4.3.

It can be seen from the table (Table 4.3) that, as the composition of the transition metal (dopant) increases, the mid bond densities of the diluted magnetic semiconductor along the [111] direction increases. The atomic number of vanadium (V), manganese (Mn) and cobalt (Co) are 23, 25 and 27 respectively. When the host lattice of germanium (Ge) having an atomic number 32 is doped with the element having a lesser atomic number, the mid bond density is higher. Also, when the host lattice is doped with the element having a higher atomic number, the mid bond density is lower.

Materials Research Forum LLC
doi: http://dx.doi.org/10.21741/9781945291777

Table 4.3 Comparison of the theoretical mid bond densities along the bonding direction [111] for the samples of $Ge_{1-x}M_x$ (M = V, Mn, Co).

Composition (x)	Bond length (Å)	Electron density (e/Å³)		
		$Ge_{1-x}V_x$	$Ge_{1-x}Mn_x$	$Ge_{1-x}Co_x$
0.02	1.2353	0.5485	0.5444	0.5414
0.04	1.2353	0.5614	0.5541	0.5476
0.06	1.2353	0.5744	0.5633	0.5538
0.08	1.2353	0.5876	0.5726	0.5599
0.10	1.2353	0.6009	0.5814	0.5661

4.3 Conclusion

The electron density distribution was analyzed using the theoretical estimation of the structure factors for the DMS materials $Si_{1-x}M_x$ and $Ge_{1-x}M_x$(M = Mn, Co, V) for five different compositions of x (0.02, 0.04, 0.06, 0.08, 0.10). It was concluded that, as the concentration of the dopant increases, the density of the sharing electrons decrease in the host lattice of silicon (Si) whereas it increases in the host lattice of germanium (Ge). It was also noted that, when the atomic number of the dopant is high, the sharing of electrons for the bonding becomes less in Si host lattice whereas the mid bond density is high in Ge host lattice. These results give an expectation about the distribution of the charges in the systems of the diluted magnetic semiconductors that are to be analyzed practically. There may be variations in the charge density distribution due to the impact of the experimental parameters but these results can be taken as the ideal case and may be compared with the experimental results.

References

[1]   Collins D.M. Nature, 298, 49(1982). https://doi.org/10.1038/298049a0

[2]   Doyle P.A., Turner P.S., Acta. Cryst. 24, 390 (1968).
        https://doi.org/10.1107/S0567739468000756

[3]   International Tables for Crystallography, Vol.C, Edited by A.J.C. Wilson, Dordrecht:Kluwer Academic Publishers, 1992.

[4]   Izumi F., Dilanian R.A., Recent Research Developments in Physics, Vol. 3, Part II, Transworld Research Network, Trivandrum, 2002.

[5]   Momma K., Izumi F., J. Appl. Crystallogr., 44, 1272(2011).
https://doi.org/10.1107/S0021889811038970

[6]   www.saraxraygroup.net, FORTRAN program 'sfac332' written by Dr. R.
Saravanan.

[7]   Wyckoff R.W.G., Crystal Structures, Inter-Science Publishers, Vol. I., London,
1963.

Dilute Magnetic Semiconducting (DMS) Materials, R. Saravanan   Materials Research Forum LLC
Materials Research Foundations **35** (2018)   doi: http://dx.doi.org/10.21741/9781945291777

Chapter 5

Experimental Charge Density Distribution in Prepared DMS Materials

Abstract

Chapter 5 deals with the estimation and the analysis of the experimental charge density of the prepared materials which are derived from the structure factors obtained from the observed powder X-ray diffraction intensities. The quantitative analysis of the estimated experimental charge density distribution of the grown DMS materials is done using the mapping of charge density in 3-dimensional, 2-dimensional and 1-dimensional spaces.

Keywords

Experimental Charge Density, DMS Materials, Structure Factors, Powder X-ray, Intensity, 3D Charge Density, 2D Charge Density, 1D Charge Density

Contents

5.1 Introduction

The analysis of electronic charge density distribution is unavoidable in the characterization of materials since the physical and electronic properties of the materials depend on the distribution of the electron charges.

In order to understand the electronic properties of a material, an accurate electron density distribution is essential which is obtained in this work by the maximum entropy method (MEM) (Collins, 1982). In this research work, the electron density calculation using MEM has been accomplished using the software PRIMA (Izumi et al., 2002). The derived electron density distribution has been visualized by employing the visualizing software VESTA (Momma et al., 2011). The three dimensional electron density distribution mapping enables us to visualize the actual distribution of the charges in the unit cell. The two dimensional electron density distribution mapping can visualize the amount of charges distributed on a particular two dimensional plane. The one dimensional electron density profile enables us to quantify the charge density between any two given atoms. In this chapter, the electron density distribution of the synthesized Si and Ge based diluted magnetic semiconductors is discussed in detail based on the 3D, 2D and 1D electron densities.

5.2 Charge density analysis of $Ge_{1-x}Mn_x$ (x = 0, 0.04, 0.06, 0.10)

The structural parameters refined from the Rietveld (Rietveld, 1969) refinement method along with the extracted observed phase factors have been used to estimate the electronic charge density distribution. Maximum entropy method (MEM) (Collins, 1982) has been used to estimate the electronic charge density distribution. In the calculation of charge density for $Ge_{1-x}Mn_x$ (x = 0, 0.04, 0.06, 0.10), the unit cell of the system has been divided into 64 x 64 x 64 pixels along each crystallographic axis. The prior electron density at each pixel has been fixed uniformly as F_{000}/a_0^3, where F_{000} is the total number of electrons in the unit cell and a_0 is the cell parameter of the $Ge_{1-x}Mn_x$ system. The Lagrange parameter was chosen so that, while maximizing the entropy, the criterion C becomes unity after a minimum number of iterations. The quantitative enumeration of the charge density using MEM has been done using the software PRIMA (Izumi et al., 2002). The mapping of the charge density distribution has been done using a visualizing software VESTA (Momma et al., 2011). The parameters refined using the MEM refinement technique are given in table 5.1.

The obtained three dimensional electron density distribution for the system of $Ge_{1-x}Mn_x$ has been mapped at same isosurface levels in order to visualize and compare the charge density at the valence region. The 3D charge density maps for the system of $Ge_{1-x}Mn_x$

are shown in figures 5.1 (a) to (d). The valency and atomic number of Mn is lower than Ge but the covalent radius of Mn is high (1.39Å for low spin and 1.61Å for high spin, Haynes, 2014). Hence, the charge density around the atomic site of Ge decreases as the Ge atoms are replaced by Mn atom in the host matrix. At the same time, doping of Mn increases the interatomic charge density, thus enhancing the metallicity of the system of $Ge_{1-x}Mn_x$. When x = 0.04, formation of intermetallic domains Ge_3Mn_5 and Ge_8Mn_{11} is comparatively low. Hence, the charge density decreases at the atomic site and at the mid-bond region. As the dopant concentration is increased in the host lattice, the formation of the intermetallic domains and Andersons mixing effect (Weng et al., 2005) between the *d* orbital and the delocalized *s* orbital of Ge, make the interatomic region to be clouded with localized charges. The charge density in the mid-bond region for 4% dopant concentration (Figure 5.1 (b)) is lesser than of 6% dopant concentration (Figure 5.1 (c)). This evidences the formation of intermetallic domains in the system which also introduces half metal like behavior in the system. The lack of interatomic charges in the system having 10% dopant concentration confirms the fact that, the dilution of the dopant Mn in the host lattice is minimum. This authenticates the critical concentration of obtaining the room temperature ferromagnetic $Ge_{1-x}Mn_x$ to be around x = 0.06. This result has been supported by the theoretical analysis made by Weng et al. (2005) in which the critical limit was found to be x = 0.0625. It has been inferred here that, when the concentration of the dopant (x) is increased beyond 0.06, the Mn-rich phases dominates and disorder is introduced in $Ge_{1-x}Mn_x$.

Table 5.1 Parameters from MEM refinement for $Ge_{1-x}Mn_x$.

Parameter	Ge	$Ge_{0.96}Mn_{0.04}$	$Ge_{0.94}Mn_{0.06}$	$Ge_{0.90}Mn_{0.10}$
Number of cycles	1767	271	282	272
Prior density, $\tau(r_i)$ (e/Å^3)	1.4150	1.4106	1.3897	1.3755
Resolution (Å/pixel)	0.0884	0.0882	0.0886	0.0885
Lagrange parameter (λ)	0.0028	0.0185	0.0191	0.0186
R_{MEM} (%)	0.9512	0.9785	0.9802	0.9841
wR_{MEM} (%)	1.2083	1.2069	1.3648	1.3658

R_{MEM}-Reliability index for the MEM refinement
wR_{MEM}-Weighted reliability index for the MEM refinement

(a)   **(b)**

(c)   **(d)**

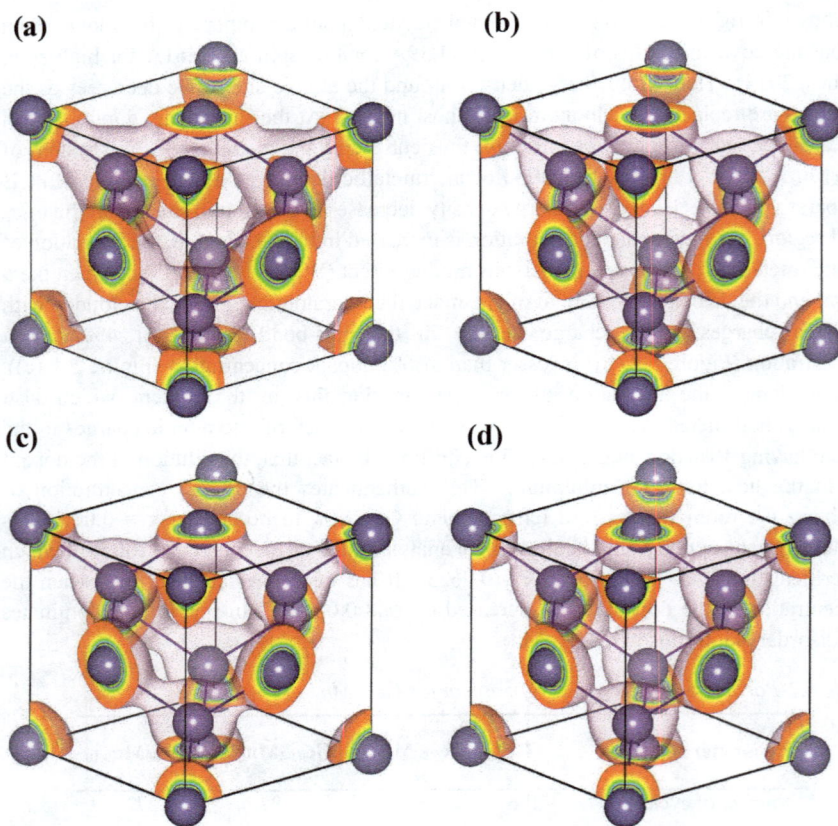

Figure 5.1 Three dimensional charge density distribution in the unit cell of (a) Ge (b) $Ge_{0.96}Mn_{0.04}$ (c) $Ge_{0.94}Mn_{0.06}$ (d) $Ge_{0.90}Mn_{0.10}$ (Isosurface level: 0.53 e/$Å^3$).

The two dimensional charge density distribution for the systems of $Ge_{1-x}Mn_x$ (x = 0, 0.04, 0.06, 0.10) on the plane (110) has been mapped with the contour level between 0.05 e/$Å^3$ and 1.0 e/$Å^3$ with the interval of 0.03 e/$Å^3$ and are shown in figures 5.2 (a) to (d). These figures show more detailed arrangement of charges between the atoms. Figure 5.2 (c) shows the presence of hybridized orbitals and resultant accumulation of charges in the valence region. The charge distribution mapped on the plane (100) with the contour level between 0.05 e/$Å^3$ and 1.0 e/$Å^3$ with the interval of 0.03 e/$Å^3$ are shown in figures 5.3 (a) to (d). The figures show exactly how the host atom Ge gets more charges when x = 0.06.

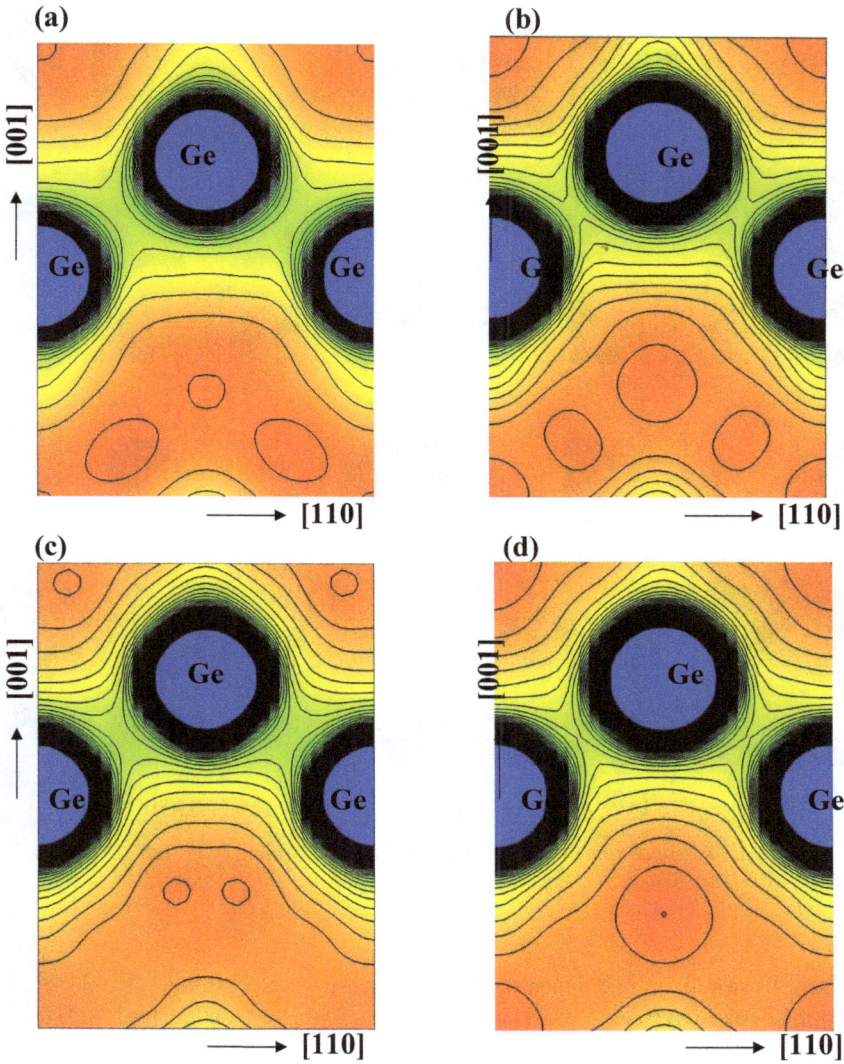

Figure 5.2 Two dimensional charge density distribution on the plane (110) for (a) Ge (b) $Ge_{0.96}Mn_{0.04}$ (c) $Ge_{0.94}Mn_{0.06}$ (d) $Ge_{0.90}Mn_{0.10}$ (Contour range: 0.05 $e/Å^3$ to 1.0 $e/Å^3$; Contour interval: 0.03 $e/Å^3$).

Dilute Magnetic Semiconducting (DMS) Materials, R. Saravanan   Materials Research Forum LLC
Materials Research Foundations **35** (2018)   doi: http://dx.doi.org/10.21741/9781945291777

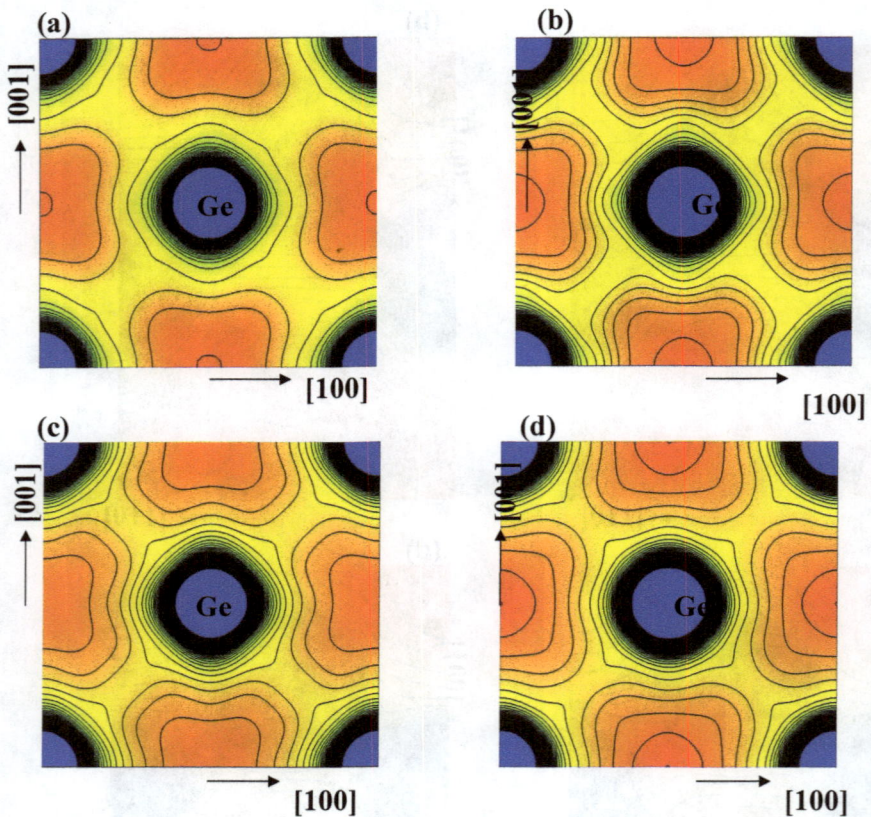

*Figure 5.3 Two dimensional charge density distribution on the plane (100) for (a) Ge (b)
$Ge_{0.96}Mn_{0.04}$ (c) $Ge_{0.94}Mn_{0.06}$ (d) $Ge_{0.90}Mn_{0.10}$ (Contour range: 0.05 e/$Å^3$ to 1.0 e/$Å^3$;
Contour interval: 0.03 e/$Å^3$).*

To quantify these results, one dimensional charge density profiles have been plotted
along the three basic directions [100], [110] and [111] and are shown in figures 5.4 (a) to
(c). The numerical values of the charge density obtained from the MEM analysis along
the three directions are tabulated in table 5.2. The charge density observed at (3,-1) bond
critical point (BCP) (Weng et al., 2005) quantifies the presence of the localized charges at
the mid-bond region (Table 5.2). The intermetallic domains and hybridization of orbitals

make the charge density at the bond critical point [111] to be $0.340e/Å^3$ when x = 0.06. But, the mid bond density is less for the concentrations of x = 0.04 and x = 0.10. These results are compounded with the expansion and compression in the lattice to allow the enhancement of the ferromagnetic order in the chosen system. Also, a unique ordering of spatial charges does happen in the chosen system of $Ge_{1-x}Mn_x$ which makes it to behave magnetically.

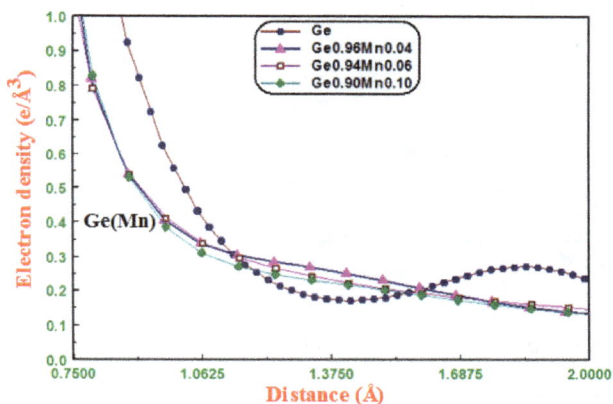

Figure 5.4 (a) One dimensional electron density profiles of $Ge_{1-x}Mn_x$ (x = 0, 0.04, 0.06, 0.10) along [100] direction.

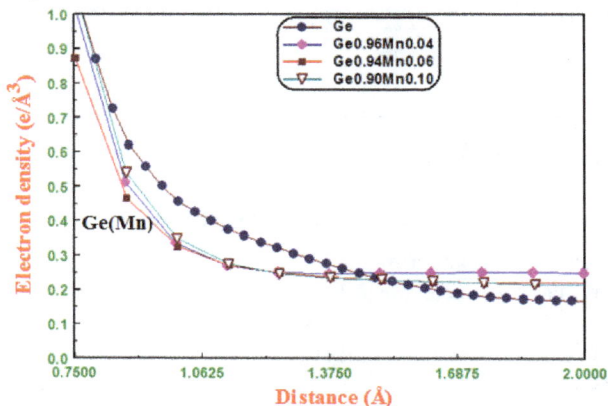

Figure 5.4 (b) One dimensional electron density profiles of $Ge_{1-x}Mn_x$ (x = 0, 0.04, 0.06, 0.10) along [110] direction.

Materials Research Forum LLC
doi: http://dx.doi.org/10.21741/9781945291777

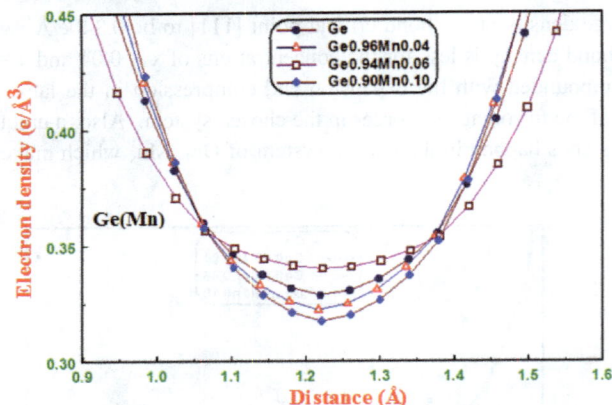

Figure 5.4 (c) One dimensional electron density profiles of $Ge_{1-x}Mn_x$ (x = 0, 0.04, 0.06, 0.10) along direction [111].

Table 5.2 Electron densities of $Ge_{1-x}Mn_x$ obtained from MEM refinement along different directions.

Direction	Ge		$Ge_{0.96}Mn_{0.04}$		$Ge_{0.94}Mn_{0.06}$		$Ge_{0.90}Mn_{0.10}$	
	r (Å)	$\rho(r)$ (e/Å³)	r (Å)	$\rho(r)$ (e/Å³)	r (Å)	$\rho(r)$ (e/Å³)	r(Å)	$\rho(r)$ (e/Å³)
[100]	1.421	0.169	2.823	0.092	2.564	0.112	2.832	0.080
[110]	2.010	0.169	1.373	0.247	1.880	0.221	2.003	0.216
[111]	1.231	0.366	1.218	0.322	1.223	0.340	1.222	0.317

r (Å)-Bond length
$\rho(r)$ (e/Å³)-Electron density at the bond length r (Å)

5.3 Charge density analysis of $Ge_{1-x}V_x$ (x = 0.03, 0.06, 0.09)

The electron density distribution of the diluted magnetic semiconductor $Ge_{1-x}V_x$ (x = 0.03, 0.06, 0.09) has been determined by maximum entropy method (MEM) (Collins, 1982) using the structure factors extracted from the experimental powder X-ray diffraction data. For the extraction of the structure factors, Rietveld refinement technique (Rietveld, 1969) was used. In this work, the unit cell was divided into 64 x 64 x 64 pixels along each crystallographic axis and a uniform prior density was fixed as F_{000}/a_0^3 where, F_{000} is the total number of electrons in the unit cell and a_0 is the cell parameter of the sytem of $Ge_{1-x}V_x$. The refinement was continued by the iterative process until the convergence criterion C = 1 was reached after minimum number of iterations. The parameters refined using MEM method for $Ge_{1-x}V_x$ have been tabulated in table 5.3. The elucidation and visualization of the electronic charge densities were done quantitatively using the software packages PRIMA (Izumi et al., 2002) and VESTA (Momma et al., 2011) respectively.

The three dimensional electronic charge densities mapped with the same isosurface levels for the samples $Ge_{0.97}V_{0.03}$, $Ge_{0.94}V_{0.06}$ and $Ge_{0.91}V_{0.09}$ are shown in figures 5.5 (a) to (c) respectively. It can be clearly seen that, the density of the bonding electrons increases as more vanadium atoms are added to the host lattice of Ge. This is because, when the paramagnetic solute like vanadium is incorporated in a metal like Ge, there is a possibility of static spin-density wave arising due to the interaction between the dopant and the host (Overhauser, 1960). As a result, readjustment of conduction electrons occurs where half of the spins will be parallel and the other half will be antiparallel. The result is a uniform flat charge density between the atoms.

The two dimensional charge density maps on the (110) plane for $Ge_{1-x}V_x$ are shown in figures 5.6 (a) to (c). The electronic charge distribution is mapped across the face (110) with the contour level between 0.02 $e/Å^3$ and 3.0 $e/Å^3$ with the interval of 0.03 $e/Å^3$. It is clear from the figures that (Figure 5.6 (a) to (c)), there is an increase in the bonding charges as the concentration of vanadium increases in the host lattice. This also causes the increase in the value of the cell parameter as the concentration of the dopant increases. The presence of the intermediate charges is expected to hold the atoms together and should strengthen the bond. But, the presence of the spin density wave with the spins arranged alternatively eliminates the possibility of tight binding and hence the atoms are allowed to vibrate with more root mean square (RMS) amplitude (B_{iso}) than before (Table 3.8).

Table 5.3 Parameters from MEM refinement for $Ge_{1-x}V_x$

Parameter	$Ge_{0.97}V_{0.03}$	$Ge_{0.94}V_{0.06}$	$Ge_{0.91}V_{0.09}$
Number of cycles	277	186	190
Prior density, $\tau(r_i)$ $(e/Å^3)$	1.3773	1.3911	1.4042
Resolution (Å/pixel)	0.0884	0.0884	0.0883
Lagrange parameter (λ)	0.0188	0.0285	0.0289
R_{MEM} (%)	0.8709	0.8464	0.8425
wR_{MEM} (%)	1.0413	1.0194	1.0145

R_{MEM}-Reliability index for the MEM refinement
wR_{MEM}-Weighted reliability index for the MEM refinement

(a)   **(b)**

(c)

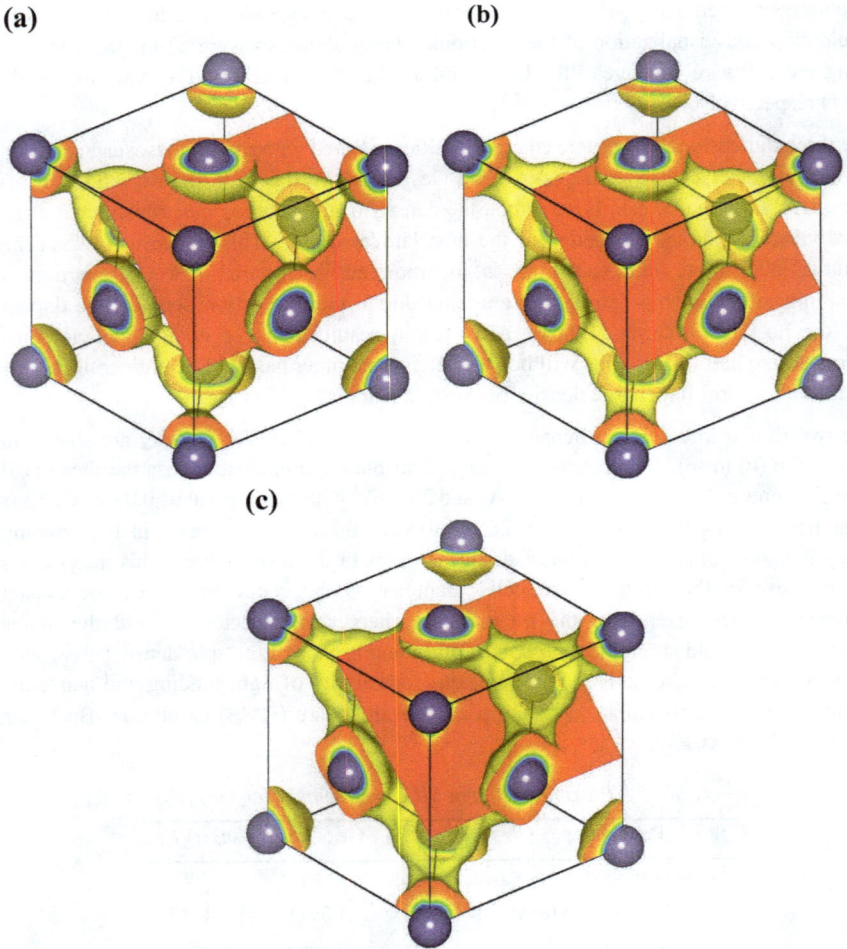

*Figure 5.5 Three dimensional charge density distribution in the unit cell of $Ge_{1-x}V_x$ with (110) plane shaded (a) $Ge_{0.97}V_{0.03}$ **(b)** $Ge_{0.94}V_{0.06}$ **(c)** $Ge_{0.91}V_{0.09}$ (Isosurface level: 0.54 $e/\AA^3$).*

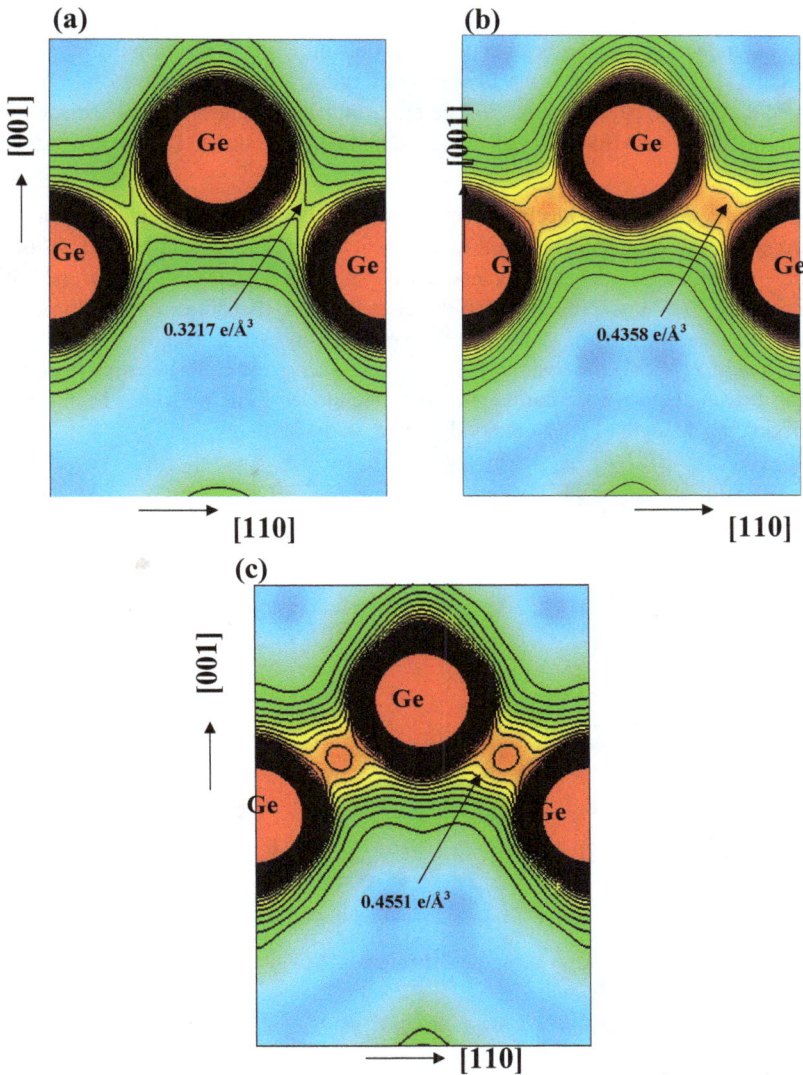

Figure 5.6 Two dimensional charge density distribution on the plane (110) for (a) $Ge_{0.97}V_{0.03}$(b) $Ge_{0.94}V_{0.06}$ (c) $Ge_{0.91}V_{0.09}$ (Contour range: 0.02 $e/Å^3$ to 3.0 $e/Å^3$; Contour interval: 0.03 $e/Å^3$).

Dilute Magnetic Semiconducting (DMS) Materials, R. Saravanan   Materials Research Forum LLC

Materials Research Foundations **35** (2018)   doi: http://dx.doi.org/10.21741/9781945291777

The analysis of the electronic charge density distribution for the samples of $Ge_{0.97}V_{0.03}$, $Ge_{0.94}V_{0.06}$ and $Ge_{0.91}V_{0.09}$ has been done on the one dimensional electron density profiles along the [111] direction. The valence contribution for the three systems has been analyzed and compared in figure 5.7. The electron density values at bond critical points are obtained from the MEM analysis along the bonding direction [111] and are presented in table 5.4.

From the one dimensional electron density profiles along the bonding direction [111] (Figure 5.7), it is clear that, the mid-bond density increases with concentration of the dopant. When the percentage of vanadium doping is $x = 0.03$, the system exhibits a perfect saddle. For $x = 0.06$ and 0.09, the charge density becomes uniform throughout the region and uplift the one dimensional charge density profile (Figure 5.7). The non nuclear maxima (NNM) observed in the mid bond is a typical characteristics of covalent bonding existing between similar atoms. The magnitude of this NNM has been used for quantifying the strength of covalent bond existing in the system.

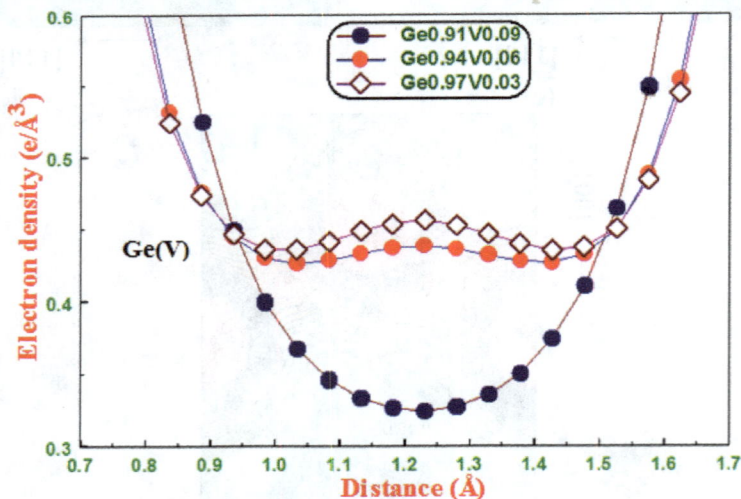

Figure 5.7 One dimensional electron density profiles of $Ge_{0.97}V_{0.03}$, $Ge_{0.94}V_{0.06}$ and $Ge_{0.91}V_{0.09}$ along [111] direction.

Dilute Magnetic Semiconducting (DMS) Materials, R. Saravanan   Materials Research Forum LLC
Materials Research Foundations 35 (2018)   doi: http://dx.doi.org/10.21741/9781945291777

Table 5.4 Electron densities of $Ge_{1-x}V_x$ at bond critical points.

System	r (Å)	$\frac{\rho(r)}{(e/Å^3)}$
$Ge_{0.97}V_{0.03}$	1.2189	0.3217
$Ge_{0.94}V_{0.06}$	1.2184	0.4358
$Ge_{0.91}V_{0.09}$	1.2180	0.4551

r (Å)-Bond length
$\rho(r)$ (e/Å^3)-Electron density at the bond length r (Å)

5.4   Charge density analysis of $Ge_{1-x}Co_x$ (x = 0.03, 0.06, 0.09)

The melt grown diluted magnetic semiconductor $Ge_{1-x}Co_x$ (x = 0.03, 0.06, 0.09) was characterized for its electron density properties which were determined by maximum entropy method (MEM) (Collins, 1982).   Maximum entropy method is a method to derive the most probable map when a set of experimental data is limited.   To estimate the electron density, the structure factors extracted using Rietveld refinement technique (Rietveld, 1969) from the experimental powder X-ray diffraction data were used.   The unit cell was divided into 64 x 64 x 64 pixels along each crystallographic axis and a uniform prior density was fixed as F_{000}/a_0^3 where F_{000} is the total number of electrons in the unit cell and a_0 is the cell parameter of the system of $Ge_{1-x}Co_x$.   The parameters refined using the MEM method are tabulated in table 5.5.   The charge density of the system of $Ge_{1-x}Co_x$ was evaluated using the software package PRIMA (Izumi et al., 2002) and visualized using the software VESTA (Momma et al., 2011).

The mapped three dimensional electronic charge densities of $Ge_{0.97}Co_{0.03}$, $Ge_{0.94}Co_{0.06}$ and $Ge_{0.91}Co_{0.09}$ are shown in figures 5.8 (a) to (c).   The three dimensional electron densities for the three samples have been mapped using the same isosurface level of 0.353 e/Å^3. From the mapping (Figures 5.8(a) and (b)), it is evident that, upto 6% of Co concentration, the midbond density increases due to the addition of the dopant Co atoms. However, when the dopant concentration is further increased to 9% (Figure 5.8(c)) the value of the mid bond density decreases since, the dopant atom is not incorporated properly in the host lattice of Ge.

The two dimensional charge density has been mapped on the plane (110) for the samples, with the contour level between 0.2 e/Å^3 and 2 e/Å^3 with an interval of 0.05 e/Å^3 in order to visualize the valence region of the bond.   The mapped two dimensional charge densities of $Ge_{1-x}Co_x$ are shown in figures 5.9 (a) to (c).   The two dimensional mapping of the samples elucidate the above said result that, for 9% of Co doping, all the dopant

Dilute Magnetic Semiconducting (DMS) Materials, R. Saravanan   Materials Research Forum LLC
Materials Research Foundations **35** (2018)   doi: http://dx.doi.org/10.21741/9781945291777

atoms have not been incorporated in the host lattice, thus reducing the value of the mid bond density. The mid bond density increases when the dopant concentration increases from 3% to 6% and then decreases when the concentration of the dopant is further increased to 9%. The mid bond density of undoped Ge is 0.366 e/Å^3 which is lower than that for 3% and 6% and higher than that for 9% of dopant concentrations.

When electron is transferred from Ge to Co, the characteristics of Co dominates Ge, resulting in the increase in the metallic behavior of the Ge:Co system. It is known that, when the charge transfer takes place between the atoms, the core of the acceptor does not accommodate the transferred electrons in the valence region preferably between the bonds due to electrostatic interactions. This results in the accumulation of charges at the bond critical point (BCP) which results in the metallic behavior of the system.

Table 5.5 Parameters from MEM refinement for Ge$_{1-x}$Co$_x$.

Parameter	Ge[*]	Ge$_{0.97}$Co$_{0.03}$	Ge$_{0.94}$Co$_{0.06}$	Ge$_{0.91}$Co$_{0.09}$
Number of cycles	1767	741	744	733
Prior density, $\tau(r_i)$ (e/Å^3)	1.4150	1.4090	1.4060	1.3927
Resolution (Å/pixel)	0.0884	0.0884	0.0883	0.0884
Lagrange parameter (λ)	0.0028	0.0084	0.0084	0.0083
R$_{MEM}$ (%)	0.9512	0.8813	0.8883	0.9051
wR$_{MEM}$ (%)	1.2083	1.1178	1.1120	1.2119

R$_{MEM}$-Reliability index for the MEM refinement
wR$_{MEM}$-Weighted reliability index for the MEM refinement
*Sheeba et al., 2012

Dilute Magnetic Semiconducting (DMS) Materials, R. Saravanan   Materials Research Forum LLC
Materials Research Foundations **35** (2018)   doi: http://dx.doi.org/10.21741/9781945291777

(a)

(b)   **(c)**

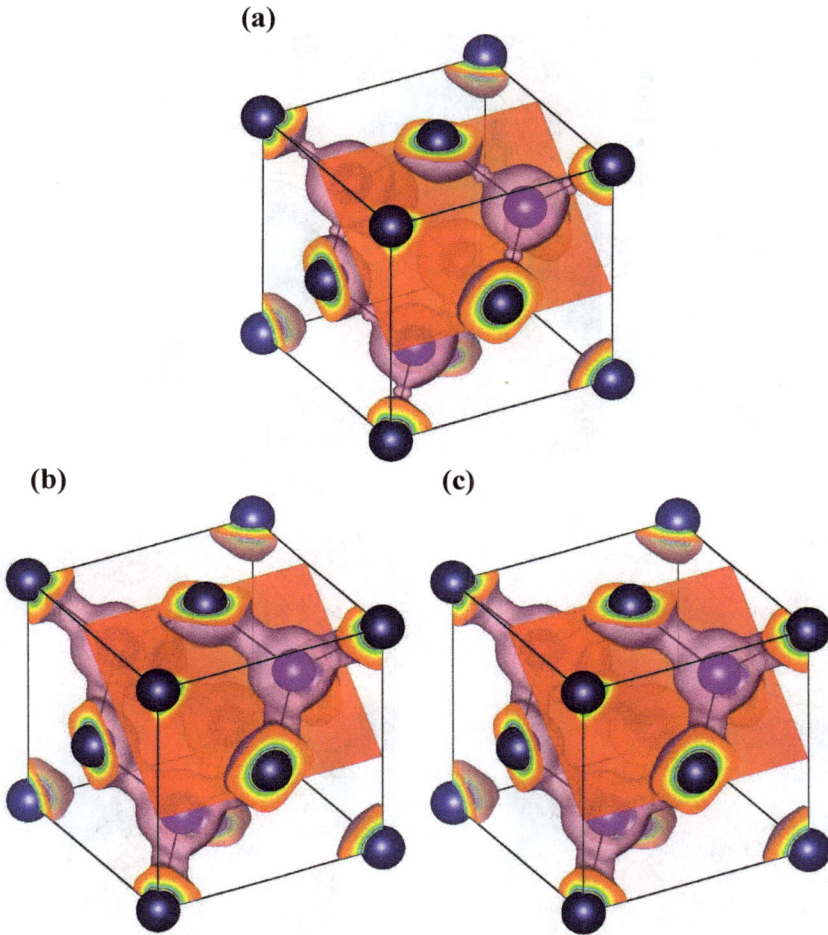

*Figure 5.8 Three dimensional charge density distribution in the unit cell of $Ge_{1-x}Co_x$ with (110) plane shaded (a) $Ge_{0.97}Co_{0.03}$ **(b)** $Ge_{0.94}Co_{0.06}$ **(c)** $Ge_{0.91}Co_{0.09}$ (Isosurface level: $0.353\ e/\text{Å}^3$).*

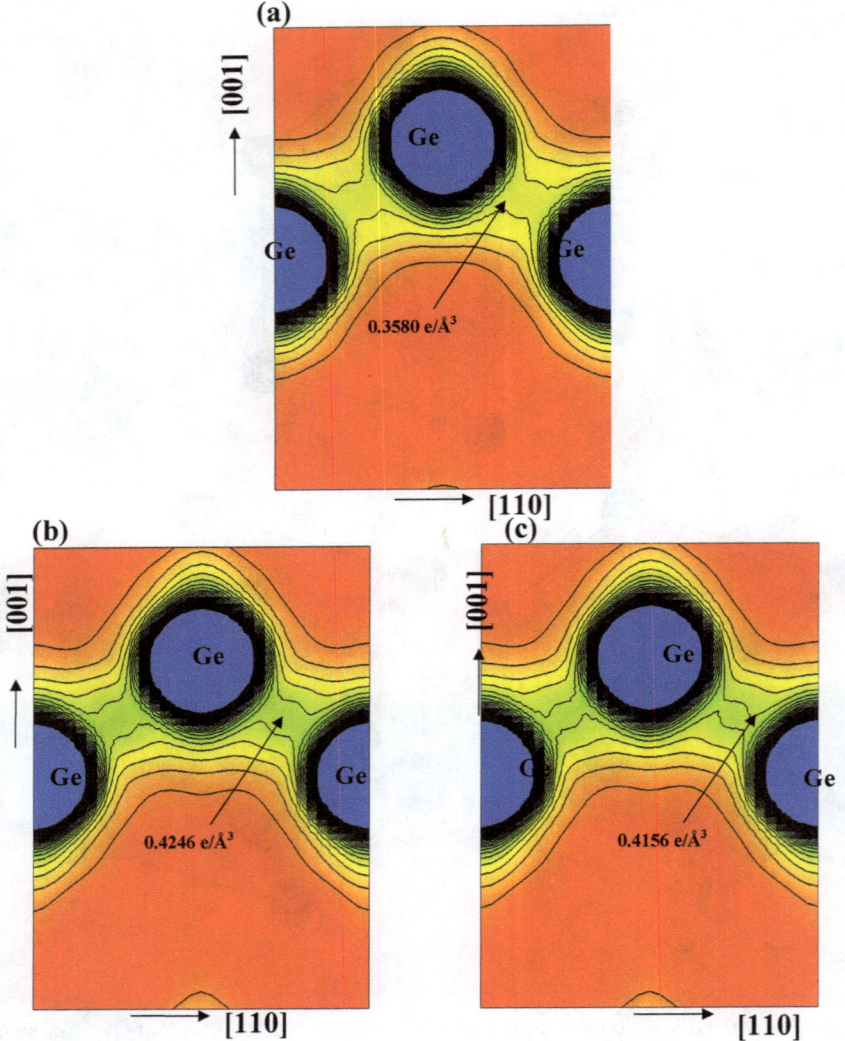

Figure 5.9 Two dimensional charge density distribution on the plane (110) for $Ge_{0.97}Co_{0.03}$ (b) $Ge_{0.94}Co_{0.06}$ (c) $Ge_{0.91}Co_{0.09}$ (Contour range: 0.2 e/Å³ to 2.0 e/Å³; Contour interval: 0.05 e/Å³).

The one dimensional electron density profiles of the samples of $Ge_{0.97}Co_{0.03}$, $Ge_{0.94}Co_{0.06}$ and $Ge_{0.91}Co_{0.09}$ are used to analyze the charge density distribution along [111] direction. The valence contributions for the three systems are shown in figure 5.10. The values of the electron densities at bond critical points obtained from the MEM analysis along the bonding direction [111] are tabulated in table 5.6. From the one dimensional electron density mapping, it is clear that, the mid bond density increases when the dopant concentration increases from 3% to 6%. Also, the mid bond density decrease when the dopant concentration is further increased to 9%.

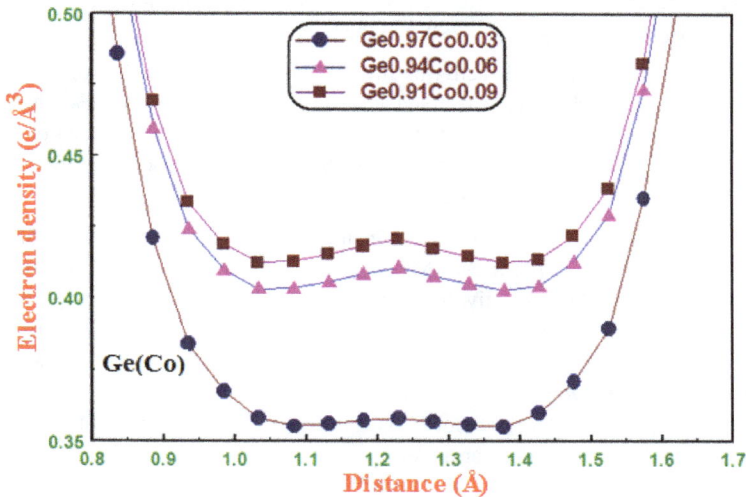

Figure 5.10 One dimensional electron density profiles of $Ge_{0.97}Co_{0.03}$, $Ge_{0.94}Co_{0.06}$ and $Ge_{0.91}Co_{0.09}$ along [111] direction.

Table 5.6 Electron densities of $Ge_{1-x}Co_x$ at bond critical points

System	r (Å)	ρ(r) (e/Å^3)
Ge*	1.231	0.366
$Ge_{0.97}Co_{0.03}$	1.2308	0.3580
$Ge_{0.94}Co_{0.06}$	1.2307	0.4246
$Ge_{0.91}Co_{0.09}$	1.2300	0.4156

r (Å)-Bond length
ρ(r) (e/Å^3)-Electron density at the bond length r (Å)
*Sheeba et al., 2012

5.5   Charge density analysis of ball milled Si and $Si_{0.98}Mn_{0.02}$

The samples of $Si_{0.98}Mn_{0.02}$ milled for 100 hrs and 200 hrs prepared in this work are termed as $Si_{0.98}Mn_{0.02}$ (100h) and $Si_{0.98}Mn_{0.02}$ (200h) respectively for convenience. The ball milled samples of $Si_{0.98}Mn_{0.02}$ (100h) and $Si_{0.98}Mn_{0.02}$ (200h) along with the undoped sample of silicon (Si) were analyzed for their charge density distributions. The charge density analysis has been done by reconstructing the charge density derived from the Rietveld refined (Rietveld, 1969) structure factors using the maximum entropy method (MEM) (Collins, 1982). The charge density distributions have been analyzed by plotting the charge density maps in three dimension and two dimension for the samples of Si, $Si_{0.98}Mn_{0.02}$ (100h) and $Si_{0.98}Mn_{0.02}$ (200h) in the bonding planes.

In this research work, the unit cell was divided into 64 x 64 x 64 pixels. The prior electron density at each pixel was fixed uniformly as F_{000}/a_0^3 where F_{000} is the total number of electrons in the unit cell and a_0 is the cell dimension of the system of $Si_{1-x}Mn_x$. The Lagrange parameter was chosen suitably so that the convergence criterion having the value unity is reached after a minimum number of iterations.

The parameters refined using MEM refinement are presented in table 5.7.   The parameters of pure Si and phase $Si_{0.98}Mn_{0.02}$ (100h) were been refined very easily under 53 and 27 iterative cycles respectively. For the deconvolution of $Si_{0.98}Mn_{0.02}$ (200h) from the diffraction profile, 707 iterative cycles were needed to get the refined charge density with weighted reliability index (wR_{MEM}) at the higher side of 7.9%. The Rietveld refinement (Rietveld, 1969) and the MEM (Collins, 1982) refinement have indicated that the chosen sample $Si_{0.98}Mn_{0.02}$ cannot be ball milled for more than 200 h. To confirm this, the sample was ball milled for about 250 h and the resultant was a powdery sample that produced a mere hump in the XRD profile which could not be used for the desired charge density analysis and hence is not presented here.

The MEM calculations were done using the software package PRIMA (Izumi et al., 2002). The visualization of the charge density of the samples was accomplished using the software package VESTA (Momma et al., 2011). The three dimensional charge density mapping using the same isosurface level of 0.43 e/Å^3 for Si, $Si_{0.98}Mn_{0.02}$ (100h) and $Si_{0.98}Mn_{0.02}$ (200h) are shown in figures 5.11 (a), (b) and (c) respectively. The three dimensional charge density in the unit cell of undoped silicon (Si) (Figure 5.11 (a)) shows high mid-bond charges along the bonding direction. As the dopant Mn^{2+} is incorporated into the system, the hole density increases and hence, the mediated charges participating in the bonding decreases. This can be clearly seen in figure 5.11 (b) which enables us to visualize the reshaped Si atom after the accumulation of the charges from the intermediate bonding region. This behavior has the tendency to increase the resistivity of the system which forms the basis for hole-mediated ferromagnetic exchange interactions in Mn doped Si (Lan Anh et al., 2009). The three dimensional structure of $Si_{0.98}Mn_{0.02}$ (200h) (Figure 5.11 (c)) shows the depletion of charges in the middle, thereby symmetrically enhancing the shape of Si at its regular lattice position. This is attributed to the increase in the hole density because of the doping of Mn. Figure 5.11(c) shows that, the generation of the secondary phase SiMn in $Si_{0.98}Mn_{0.02}$ (200h) would be more which can be authenticated from the enhancement of the intensity of (102) peak profile of SiMn among $Si_{0.98}Mn_{0.02}$ (200h) XRD profile. It can also be concluded that, ball milling for more than 200 hours will introduce greater concentration of the secondary phase and will spoil the realization of the expected diluted magnetic semiconductor system Si:Mn. Therefore, in order to realize DMS Si:Mn, the ball milling should be done below the milling time of 200 h.

Table 5.7 Parameters from MEM refinement for $Si_{1-x}Mn_x$.

Parameter	Si	$Si_{0.98}Mn_{0.02}$(100h)	$Si_{0.98}Mn_{0.02}$(200h)
Number of cycles	53	27	707
No. of electrons in the unit cell (F_{000})	112	113.76	114
Prior density, $\tau(r_i)$ (e/$Å^3$)	0.7330	0.7054	0.7040
Resolution (Å/pixel)	0.0835	0.0850	0.0852
Lagrange parameter (λ)	0.0152	0.0513	0.0227
R_{MEM} (%)	1.3764	4.0562	17.8075
wR_{MEM} (%)	2.3210	3.3129	7.9713

R_{MEM}-Reliability index for the MEM refinement
wR_{MEM}-Weighted reliability index for the MEM refinement

(a)   **(b)**

(c)

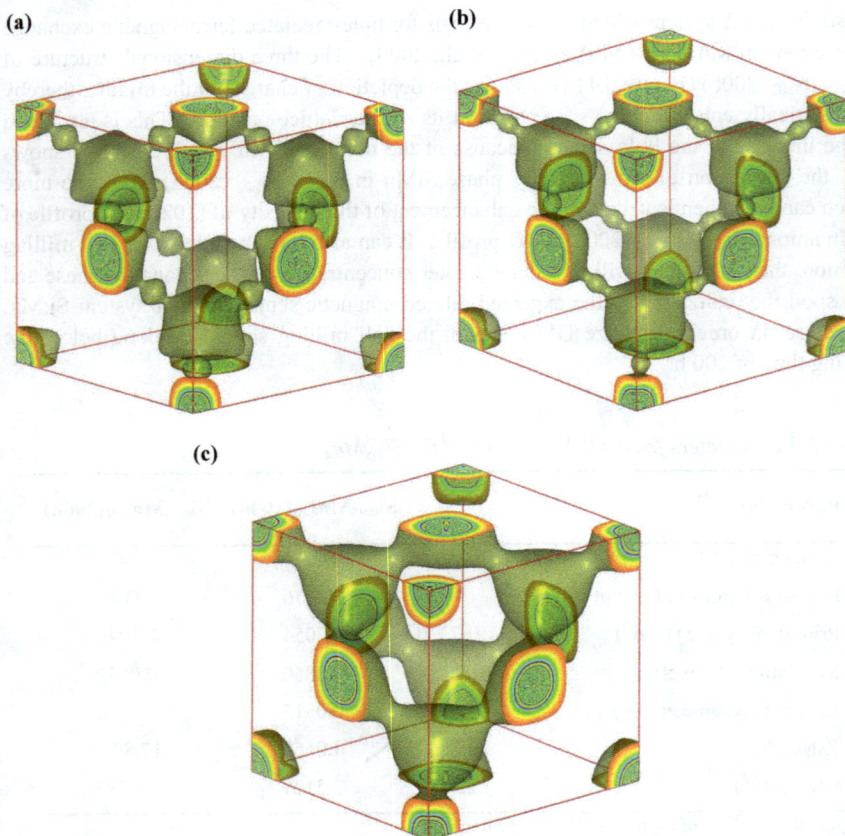

Figure 5.11 Three dimensional charge density distribution in the unit cell of (a) Si (b) $Si_{0.98}Mn_{0.02}(100h)$ (c) $Si_{0.98}Mn_{0.02}(200h)$ (Isosurface level: 0.43 $e/Å^3$).

Dilute Magnetic Semiconducting (DMS) Materials, R. Saravanan   Materials Research Forum LLC
Materials Research Foundations **35** (2018)   doi: http://dx.doi.org/10.21741/9781945291777

The two dimensional charge density mapping on the plane (100) for the samples Si, Si, $Si_{0.98}Mn_{0.02}$ (100h) and $Si_{0.98}Mn_{0.02}$ (200h) is shown in figures 5.12 (a), (b) and (c) respectively. The two dimensional charge density has been mapped for the samples choosing the contour levels from 0.09 $e/Å^3$ to 5 $e/Å^3$ with an interval of 0.03 $e/Å^3$ in order to visualize the valence region of the bond. The mapped two dimensional charge density shows the Si atom enclosing its adjacent charges due to doping of Mn and enhancing the generation of secondary phase SiMn over increased ball milling time. The two dimensional charge density mapping on the plane (110) for the chosen samples are shown in figures 5.13 (a), (b) and (c). This shows the accumulated mid bond density that takes part in the directional covalent bonding in the case of pure Si. It also exhibits depletion of charges from the mid bond with increasing ball milling time. The increase in the hole density due to the Mn doping (Lan Anh et al., 2009) and enhanced information of the secondary phase SiMn tend to reduce the charges taking part in the covalent bond. The increase in resistivity of any material is always attributed to the depletion of charges in the bonding region thereby forming closed shell interaction (Lan Anh et al., 2009). Hence, it can be concluded that, the ball milling of the sample $Si_{0.98}Mn_{0.02}$(200h) for more than 200 h can lead the system to enter into a near insulator phase which is not desired. The one dimensional charge density variations along the bonding direction is shown in figure 5.14. The quantitative values of the electronic charge density of the samples is tabulated and given in table 5.8. The one dimensional charge density values along the bonding direction also substantiates the above discussed result.

Table 5.8 Electron densities of Si, $Si_{0.98}Mn_{0.02}$(100h) and $Si_{0.98}Mn_{0.02}$(200h) along various directions in the unit cell.

Direction	Si		$Si_{0.98}Mn_{0.02}$(100h)		$Si_{0.98}Mn_{0.02}$(200h)	
	Distance (Å)	Electron density $(e/Å^3)$	Distance (Å)	Electron density $(e/Å^3)$	Distance (Å)	Electron density $(e/Å^3)$
[100]	2.6552	0.0724	2.7040	0.0879	2.7117	0.0287
[110]	2.3342	0.2262	1.9120	0.1157	1.9368	0.0747
[111]	1.1808	0.5008	1.2025	0.5005	1.1860	0.5173

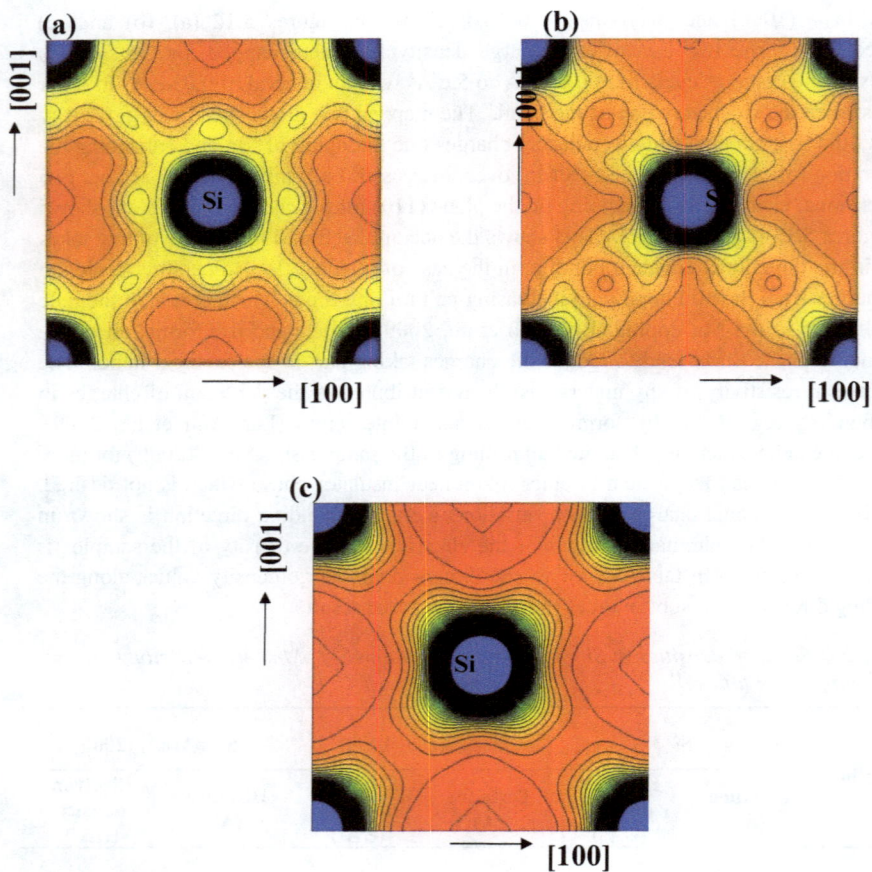

Figure 5.12 Two dimensional charge density distribution on the plane (100) for (a) Si (b) $Si_{0.98}Mn_{0.02}(100h)$ (c) $Si_{0.98}Mn_{0.02}(200h)$ (Contour range: 0.09 $e/Å^3$ to 5.0 $e/Å^3$; Contour interval: 0.03 $e/Å^3$).

Dilute Magnetic Semiconducting (DMS) Materials, R. Saravanan   Materials Research Forum LLC
Materials Research Foundations **35** (2018)   doi: http://dx.doi.org/10.21741/9781945291777

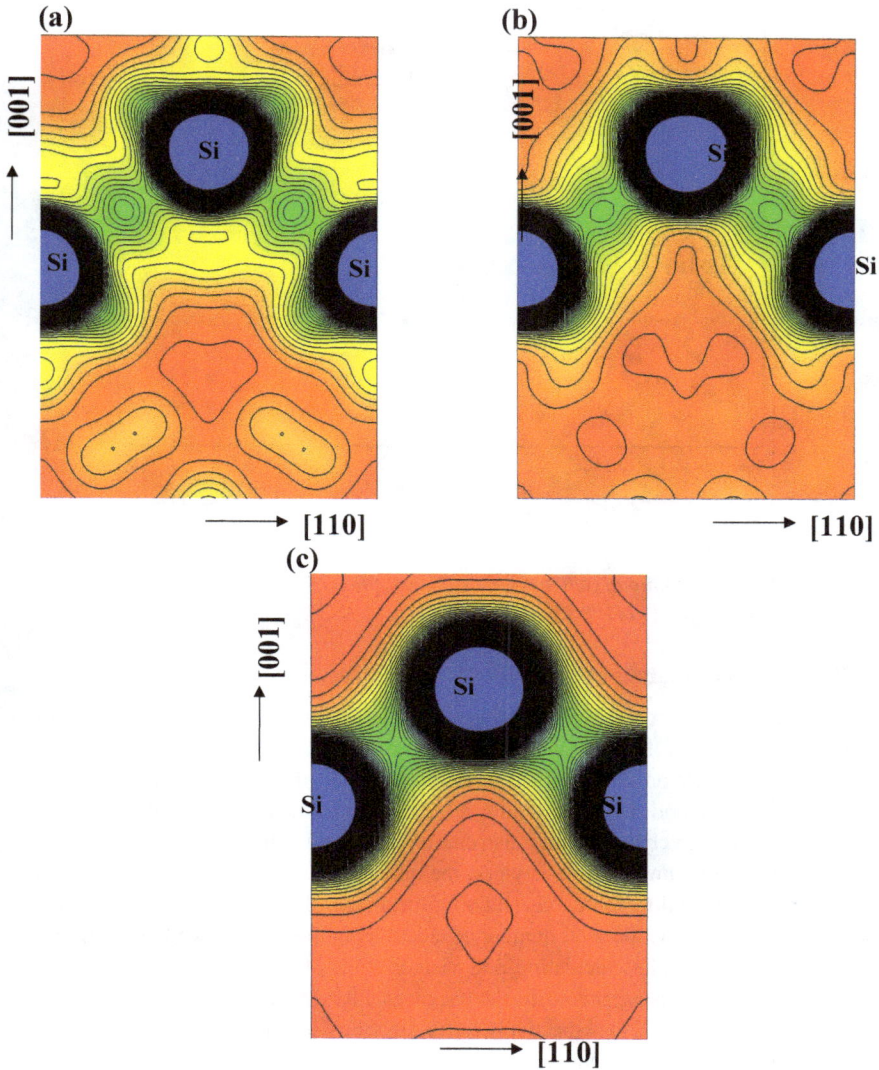

Figure 5.13 Two dimensional charge density distribution on the plane (110) for (a) Si (b) $Si_{0.98}Mn_{0.02}(100h)$ (c) $Si_{0.98}Mn_{0.02}(200h)$ (Contour range: 0.09 $e/Å^3$ to 5.0 $e/Å^3$; Contour interval: 0.03 $e/Å^3$).

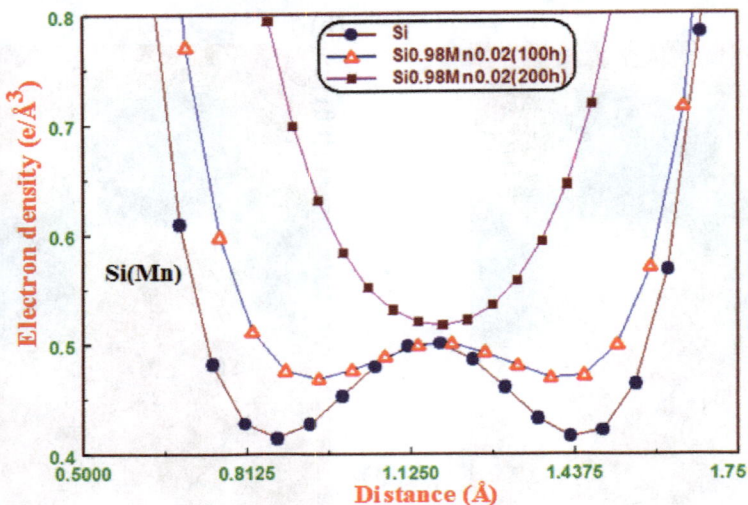

Figure 5.14 One dimensional electron density profiles of Si, $Si_{0.98}Mn_{0.02}(100h)$ and $Si_{0.98}Mn_{0.02}(200h)$ along [111] direction.

5.6   Charge density analysis of ball milled $Si_{1-x}Ni_x$

The electronic charge density distribution of the prepared DMS samples was reconstructed using the refined structure factors and their phase factors. The electronic charge density distribution of $Si_{1-x}Ni_x$ was estimated using the versatile statistical tool maximum entropy method (MEM) (Collins, 1982). Maximum entropy method is a tool that gives exact value of charge density distribution in the unit cell even when the number of the observed data is limited. In this work, the unit cell was divided into $48 \times 48 \times 48$ pixels along each crystallographic axis and a uniform prior density of F_{000}/a_0^3 was fixed in each pixel where F_{000} is the total number of electrons in the unit cell and a_0 is the cell parameter of $Si_{1-x}Ni_x$. The calculation and visualization of the experimental electron density has been done using the software packages PRIMA (Izumi et al., 2002) and VESTA (Momma et al., 2011) respectively. The reliability indices and other parameters obtained through MEM refinement are given in table 5.9.

The three dimensional charge density mapping of undoped Si and $Si_{1-x}Ni_x$ is shown in figures 5.15 (a) to (e). It is clear that, the spherical shape of the charge density of the

Dilute Magnetic Semiconducting (DMS) Materials, R. Saravanan   Materials Research Forum LLC
Materials Research Foundations **35** (2018)   doi: http://dx.doi.org/10.21741/9781945291777

atoms changes. The strength of the valence charges in the bonding region due to the presence of the dopants is clearly visualized.

The impurity atom prefers to occupy the substitutional position rather than interstitial sites which is evident from the shift in the peak at (111) (Figure 3.19(b)). Any feasible presence of micro clusters of the FCC structured Ni will suddenly alter this situation as the dopant (Ni) concentration is increased. It is evident from the 3-dimensional map (Figure 5.15) that, as the concentration of Ni is increased, the overlapping of the valence charges is increased and well pronounced at 9% of dopant concentration (Figure 5.15(d)) and then attains a maximum at 12% (Figure 5.15(e)). The S-M transition requires mid bond charge density to be delocalized and is to spread over the entire valence region to mimic overlapping of charges (Collver, 1977). In the prepared samples it is found that, the mid bond electron density increases as the dopant is introduced in the host system. The value of the mid bond electron density is maximum when $x = 0.12$. The values of the mid bond electron density between Si atoms in $Si_{1-x}Ni_x$ are tabulated in table 5.10.

The cross section of charge density at (110) plane is shown in figures 5.16 (a) to (e). The two dimensional charge density maps clearly reveal the increase in the mid bond charge density at bond critical point (BCP) along the bond path and when $x = 0.12$, the mid bond density is maximum. To quantify these results, one dimensional charge density profiles were plotted along the bond path and are shown in figure 5.17. This figure reveals that, as the concentration of the dopant is increased, there is a gradual increase in the BCP charge density. Interestingly, the position of the BCP has not changed very much except when $x = 0.12$ and this variation in the position of BCP may be attributed to possible influence of secondary phase, which has not affected the system until $x = 0.12$. Thus, it can also be authenticated that the systems for which $x < 0.12$ have the dopant Ni atoms substituting the host Si in the matrix rather than being at the interstitial site.

When the system makes a transition from p to n type semiconductor with the increase of dopant concentration, the impurity levels generated by the dopant Ni are prevented to contribute to the core (Collver, 1977). This infers that, charges from Ni are allowed to be accommodated only along the valence region which increases the charge density at (3,-1) bond critical point.

The electron beam evaporated films of $Si_{1-x}Ni_x$ (Collver, 1977) and the ball milled $Si_{1-x}Ni_x$ (present work) have enacted similar properties and hence both these methods of preparation can be used for preparing $Si_{1-x}Ni_x$ diluted magnetic semiconductor system which can be used for spintronic applications. However, to get rid of the secondary phase of Ni and any possible substitutional impurity, zone melting technique can be adopted.

Table 5.9 Parameters from MEM refinement for $Si_{1-x}Ni_x$.

Parameter	Si	$Si_{0.97}Ni_{0.03}$	$Si_{0.94}Ni_{0.06}$	$Si_{0.91}Ni_{0.09}$	$Si_{0.88}Ni_{0.12}$
Number of cycles	113	25	24	24	32
Prior density, $\tau(r_i)$ (e/Å^3)	0.6968	0.7208	0.7509	0.7734	0.7978
Resolution (Å/pixel)	0.0850	0.0847	0.0846	0.0844	0.0842
Lagrange parameter (λ)	0.0212	0.0620	0.0615	0.0565	0.0655
R_{MEM} (%)	2.34	3.85	3.13	2.05	3.98
wR_{MEM} (%)	2.98	3.54	3.46	2.63	3.65

R_{MEM}-Reliability index for the MEM refinement
wR_{MEM}-Weighted reliability index for the MEM refinement

Table 5.10 Electron densities of $Si_{1-x}Ni_x$ at bond critical points.

System	r (Å)	$\rho(r)$ (e/Å^3)
Si	1.1702	0.2415(2)
$Si_{0.97}Ni_{0.03}$	1.1739	0.3193(5)
$Si_{0.94}Ni_{0.06}$	1.1709	0.3690(1)
$Si_{0.91}Ni_{0.09}$	1.1711	0.4087(2)
$Si_{0.88}Ni_{0.12}$	1.1805	0.4678(4)

r (Å)-Bond length
$\rho(r)$ (e/Å^3)-Electron density at the bond length r (Å)

Dilute Magnetic Semiconducting (DMS) Materials, R. Saravanan   Materials Research Forum LLC
Materials Research Foundations **35** (2018)   doi: http://dx.doi.org/10.21741/9781945291777

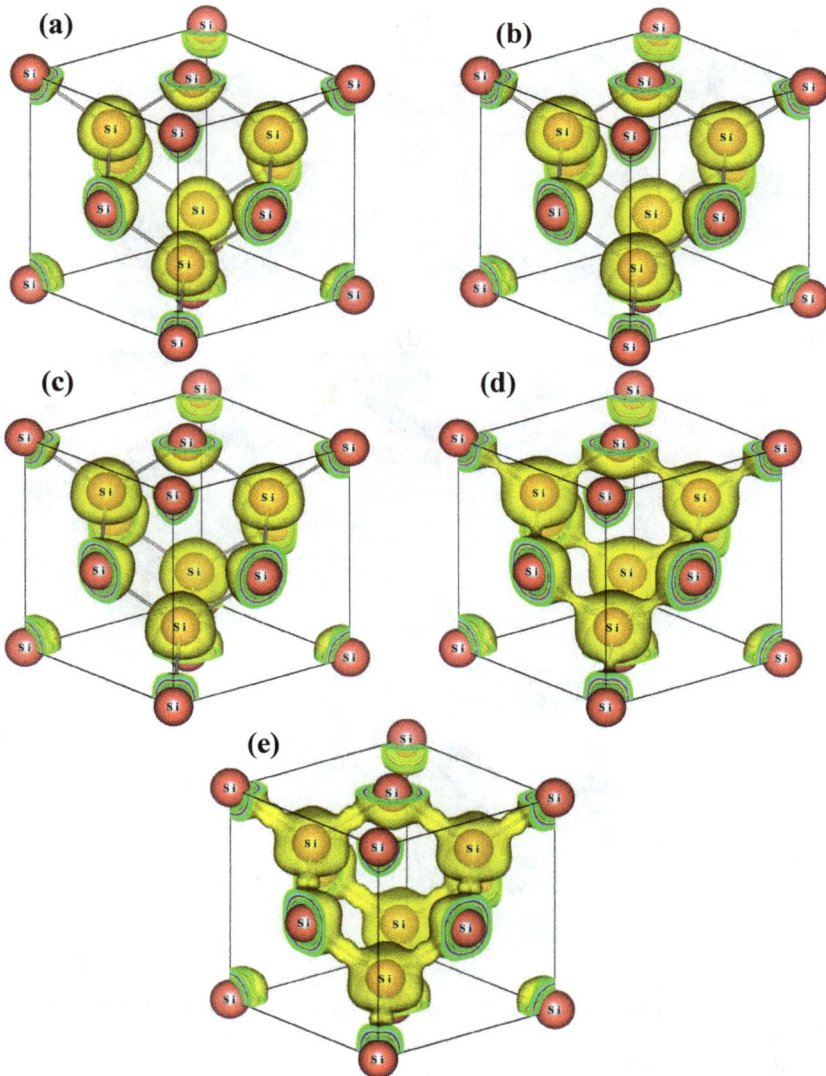

Figure 5.15 Three dimensional charge density distribution in the unit cell of (a) Si (b) $Si_{0.97}Ni_{0.03}$ (c) $Si_{0.94}Ni_{0.06}$ (d) $Si_{0.91}Ni_{0.09}$ (e) $Si_{0.88}Ni_{0.12}$ (Isosurface level: 1.5 e/$Å^3$).

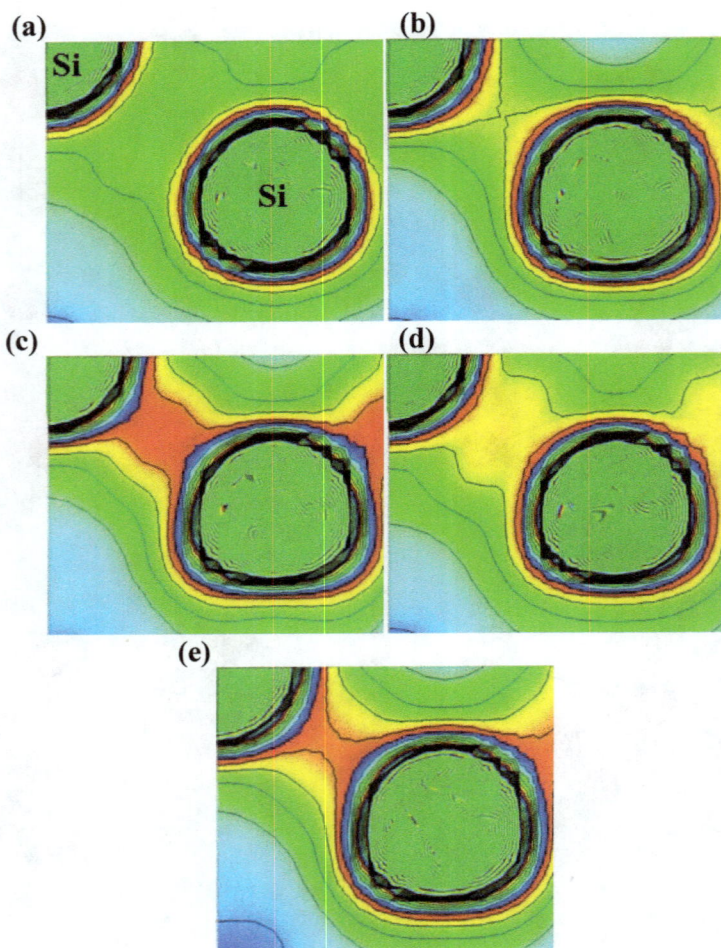

Figure 5.16 Two dimensional charge density distribution on the plane (110) for (a) Si (b) $Si_{0.97}Ni_{0.03}$ (c) $Si_{0.94}Ni_{0.06}$ (d) $Si_{0.91}Ni_{0.09}$ (e) $Si_{0.88}Ni_{0.12}$ (Contour range: 0 $e/Å^3$ to 1 $e/Å^3$; Contour interval: 0.07 $e/Å^3$).

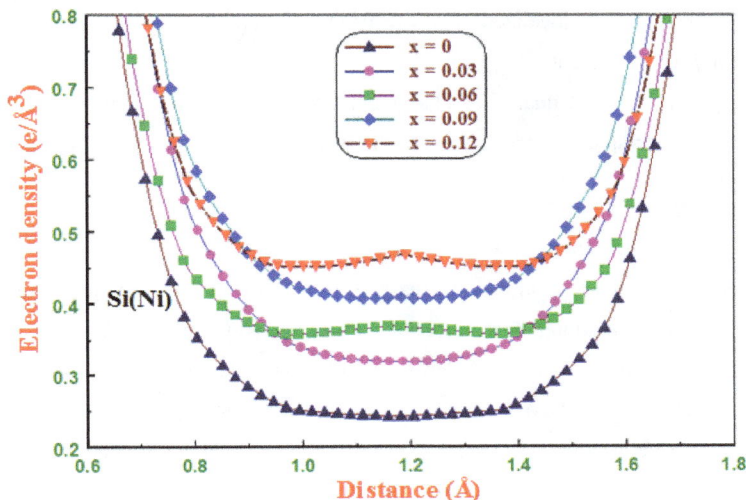

Figure 5.17 One dimensional electron density profiles of $Si_{1-x}Ni_x$ (x = 0, 0.03, 0.06, 0.09, 0.12) along [111] direction.

5.7   Conclusion

The charge density analysis for the diluted magnetic semiconducting materials of $Ge_{1-x}Mn_x$ (x = 0, 0.04, 0.06, 0.10), $Ge_{1-x}V_x$ (x = 0.03, 0.06, 0.09), $Ge_{1-x}Co_x$ (x = 0.03, 0.06, 0.09) prepared using melt technique and Si, $Si_{0.98}Mn_{0.02}$ (100h and 200h) and $Si_{1-x}Ni_x$ (x = 0, 0.03, 0.06, 0.09, 0.12) prepared using mechanical alloying has been done using the maximum entropy method. The electron density distribution of the materials has been analyzed with the help of the three dimensional, two dimensional and one dimensional charge density mapping.

5.7.1   Melt grown $Ge_{1-x}Mn_x$

i.   The charge density at the bond critical point (BCP) along the bonding direction [111] is higher when the dopant concentration is x = 0.06.

ii.   The value of the mid-bond density at x = 0.06 is higher than that of the samples having the dopant concentrations x = 0.04 and x = 0.10.

iii.   The presence of localized charges in the mid bond region is confirmed and the bonding is covalent.

iv.   The results of the quantified charge density enables us to conclude that, a unique ordering of spatial charges occur in the system of $Ge_{1-x}Mn_x$.

5.7.2   Melt grown $Ge_{1-x}V_x$

i.   The electronic charge density in the mid bond region between the atoms increases as the dopant concentration increases which can be visualized clearly in the two dimensional mapping of the electronic charge density of the system.

ii.   The one dimensional charge density profile along the bonding direction [111] also substantiates the above result.

iii.   The perfect saddle at $x = 0.03$ indicates the absence of any attracting spin density wave while at $x = 0.06$ and 0.09, the presence of the spin density wave causes the charge density to become uniform throughout the region and uplift the one dimensional electron density profile.

iv.   Covalent bonding existing between the atoms and the strength of the bond is determined by the non nuclear maxima (NNM) observed in the mid bond region.

5.7.3   Melt grown $Ge_{1-x}Co_x$

i.   The addition of the cobalt (Co) atoms in the host Ge lattice increases the mid bond electron density up to $x = 0.06$ of the dopant concentration and decreases when the dopant concentration is $x = 0.09$.

ii.   During the process of doping, charge transfer takes place between Ge and Co, Co metal dominates Ge resulting in the increase in the metallic behavior of the Ge:Co system.

iii.   Covalent bonding is confirmed in the system by the values of the mid bond electron charge density.

5.7.4 Ball milled $Si_{1-x}Mn_x$

i.   The Si atom encloses its adjacent charges due to Mn doping when the milling time is increased.

ii.   In undoped Si, accumulated mid bond charges form the covalent bonding whereas, depletion of charges from the mid bond has been observed in the ball milled samples of $Si_{0.98}Mn_{0.02}$ (100h) and $Si_{0.98}Mn_{0.02}$ (200h).

iii. Increase in the hole density due to Mn doping tends to decrease the amount of charges taking part in the covalent bonding thus enhancing a closed shell interaction which increases the resistivity in the system.

5.7.5 Ball milled $Si_{1-x}Ni_x$

i.   The impurity atoms prefer to be on the substitutional position rather than at the interstitial sites.

ii.   Overlapping of charges has been observed in the Si:Ni system as the dopant Ni concentration is increased. This is required for the S-M transition to take place in the prepared DMS materials. The overlapping of valence charges is well pronounced at $x = 0.09$ and enhances to a maximum at $x = 0.12$ of the dopant concentration.

iii. The mid bond electronic charge density increases at bond critical point (BCP) and reaches a maximum when $x = 0.12$. The position of the BCP doesn't change much except for the dopant concentration of $x = 0.12$. Hence, when $x < 0.12$, the dopant atoms substitute the host Si rather being at the interstitial site.

References

[1]   Collins D.M. Nature, 298, 49 (1982). https://doi.org/10.1038/298049a0

[2]   Collver M.M., Solid State Communications, 23, 333 (1977).
       https://doi.org/10.1016/0038-1098(77)91340-0

[3]   Haynes W.M., CRC Handbook of Chemistry and Physics, CRC Press/Taylor and
       Francis, Boca Raton, FL, 95th Edition, 2014.

[4]   Izumi F., Dilanian R.A., Recent Research Developments in Physics, Vol. 3, Part II, Transworld Research Network, Trivandrum, 2002.

[5]   Lan Anh T.T., Yu S.S., Ihm Y.E., Kim D.J., Kim H.J., Hong S.K., Kim C.S., Physica B 404, 1686 (2009). https://doi.org/10.1016/j.physb.2009.02.001

[6]   Momma K., Izumi F., J. Appl. Crystallogr., 44, 1272 (2011). https://doi.org/10.1107/S0021889811038970

[7]   Overhauser A.W., J. Phys. Chem. Solids., 13, 71 (1960). https://doi.org/10.1016/0022-3697(60)90128-1

[8]   Rietveld H.M., Journal of Applied Crystallography, 2, 65 (1969). https://doi.org/10.1107/S0021889869006558

[9]   Sheeba R.A.J.R., Saravanan R., Berchmans L.J., Materials Science is Semiconductor Processing, 15, 731 (2012). https://doi.org/10.1016/j.mssp.2012.03.007

[10]   Weng H., Dong J., Physical Review B., 71, 035201 (2005). https://doi.org/10.1103/PhysRevB.71.035201

Dilute Magnetic Semiconducting (DMS) Materials, R. Saravanan   Materials Research Forum LLC
Materials Research Foundations **35** (2018)   doi: http://dx.doi.org/10.21741/9781945291777

Chapter 6

Magnetic Properties of Prepared DMS Materials

Abstract

Chapter 6 gives the results of the analysis of the magnetic measurements of the prepared samples carried out using a vibrating sample magnetometer. This chapter also gives the correlation of the observed magnetic properties to the structural properties and the distribution of charge density of the prepared DMS materials.

Keywords

Magnetic, Hysteresis, Structural Properties, Charge Density, Ferromagnetic, Diamagnetic, Saturation Magnetization

Contents

6.1   Introduction

The analysis of the magnetic properties of the grown diluted magnetic semiconductor materials is important because, it substantiates the proper incorporation of the transition metal dopant in the host semiconductor materials. Also, it gives us an insight into the magnetic nature of the prepared material. The main aim of fabrication of the diluted magnetic semiconductors is to achieve the property of room temperature ferromagnetism in the fabricated material. In order to achieve this task, it is important to analyze the fabricated diluted magnetic semiconductor material for its magnetic properties so that, we can come to the conclusion about the conditions through which the room temperature ferromagnetism can be obtained in the fabricated materials.

In this chapter, the magnetic properties of the grown Si and Ge based diluted magnetic semiconductors have been analyzed using the magnetic measurements carried out using a vibrating sample magnetometer (VSM) and the results have been discussed.

6.2   Magnetic properties of $Ge_{1-x}Mn_x$

The magnetic measurements for the grown samples of $Ge_{1-x}Mn_x$ (x = 0.04, 0.06, 0.10) were recorded using a vibrating sample magnetometer (VSM) (Lake Shore Make, Model 7407) at the Central Instrumentation Facility (CIF), Pondicherry University, Pondicherry, India, up to a magnetic field of 10 kG at a constant temperature of 295 K. The magnetic hysteresis behavior of $Ge_{1-x}Mn_x$ is plotted in figure 6.1.

The magnetic hysteresis curve of $Ge_{1-x}Mn_x$ exhibits antiferromagnetic behavior for the dopant concentration, x = 0.04. As the magnetic field is increased, the magnetization of the samples attains a maximum and then decreases. This explains the antiferromagnetic behavior of the sample for x = 0.04. When x = 0.06, the sample transits into a perfect ferromagnet. The magnetization of the sample attains a perfect saturation as the magnetic field is increased which explains the ferromagnetic nature of the sample when x = 0.06. When x = 0.10, the system behaves like a soft ferromagnet (See the insert of Figure 6.1). The magnetization in the sample attains a maximum but is not saturated. Hence, $Ge_{1-x}Mn_x$ behaves as a soft ferromagnet when x = 0.10. The values of the coercivity and other parameters of the magnetic measurements are presented in table 6.1.

Table 6.1 Magnetic parameters of $Ge_{1-x}Mn_x$.

System	Coercivity (G)	Magnetization (emu/g) $(x10^{-3})$	Retentivity (emu/g) $(x10^{-3})$
$Ge_{0.96}Mn_{0.04}$	699.28	18.777	11.992
$Ge_{0.94}Mn_{0.06}$	488.30	22.819	10.839
$Ge_{0.90}Mn_{0.10}$	315.00	0.8	0.220

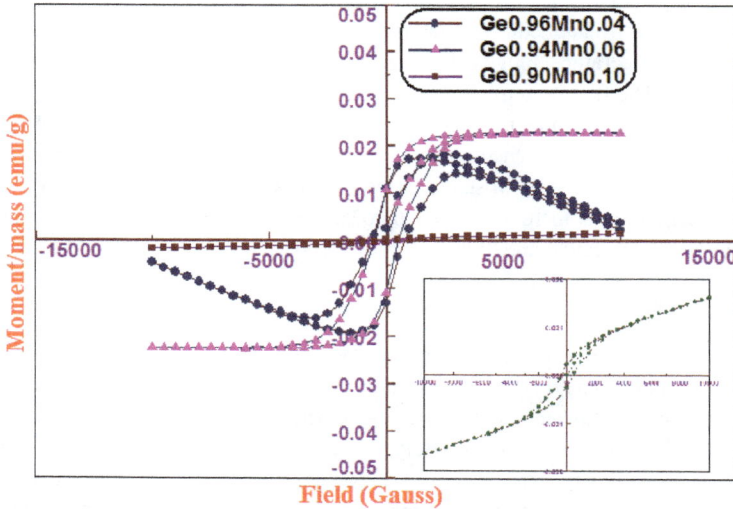

Figure 6.1 Variation of magnetic hysteresis with magnetic field in $Ge_{0.96}Mn_{0.04}$, $Ge_{0.94}Mn_{0.06}$ and $Ge_{0.90}Mn_{0.10}$ (Insert: Magnetic behavior of $Ge_{0.90}Mn_{0.10}$).

The addition of the dopant transition metal Mn into the host lattice of Ge leads to the formation of intermetallic domains $viz.$, Mn-poor Ge_8Mn_{11} initially and then to Mn-rich Ge_3Mn_5 as the dopant concentration is increased. The formation of these intermetallic compounds is the origin of ferromagnetism in the samples of $Ge_{1-x}Mn_x$ (Beigger et al., 2007). From the measurements done for the samples of $Ge_{1-x}Mn_x$, it is found that, when $x = 0.04$, the intermetallic domain Mn-poor Ge_8Mn_{11} dominates over the Mn-rich Ge_3Mn_5. The high concentration of Mn-poor Ge_8Mn_{11} leads to the competing of long range ferromagnetic order to short range antiferromagnetic order in which the antiferromagnetic behavior dominates the most. The competition between Ge_8Mn_{11} and Mn^{2+} brings antiferromagnetic behavior in the system of $Ge_{1-x}Mn_x$ when the dopant concentration is $x = 0.04$.

As the concentration of the dopant is increased to 6% ($x = 0.06$), the creation of the Mn-rich Ge_3Mn_5 dominates over its counterpart $i.e.$, Mn-poor Ge_8Mn_{11}. This leads to the perfect ferromagnetic behavior of $Ge_{0.94}Mn_{0.06}$. Based on the study on the effect of temperature on changing magnetic behavior (Beigger et al., 2007), it can be suggested that, the transition to ferromagnetic (FM) state from the antiferromagnetic (AFM) state takes place for $Ge_{1-x}Mn_x$ at 150K and this behavior is attributed to the presence of Mn-poor Ge_8Mn_{11}. $Ge_{1-x}Mn_x$ behaves as an antiferromagnet until 150K and then the transition to the ferromagnetic state takes place. $Ge_{1-x}Mn_x$ remains in the ferromagnetic state in the temperature range of 150K to 285K and from there on, it becomes paramagnetic. The first principle investigation (Weng et al., 2005) suggests that, for $x = 0.03125$, the lowest ferromagnetic energy is found in the configuration of N440 and the corresponding energy differences between the FM and the AFM states infer that, the system prefers to be in the AFM state. In their calculations (Weng et al., 2005), when the dopant concentration is increased to $x = 0.0625$, the energy difference between the AFM and FM states decreases and the system prefers to be in the FM state. From the experiment done by us, we confirm that, $Ge_{1-x}Mn_x$ behaves as predicted by the first principle calculation of Weng et al. (2005).

When the Mn content is further increased in $Ge_{1-x}Mn_x$ up to 10% ($x = 0.10$), the system tends to lose its ferromagnetic behavior and it enters into a soft magnetic state. This is due to Mn-rich Ge_3Mn_5 nanoclusters that are superparamagnetic in nature and they tend to mimic the spin-glass behavior (Beigger et al., 2007). This behavior introduces paramagnetism in the system. It is also suggested (Beigger et al., 2007) that, the Mn-poor Ge_8Mn_{11} can transit to a soft ferromagnetic state at room temperature. Hence, for $x = 0.10$, the mixture of both Mn-rich and Mn-poor domains can very well suppress the ferromagnetic behavior and induce paramagnetism in the system. All these combined effects make the system with $x = 0.10$ to be a soft ferromagnet.

Dilute Magnetic Semiconducting (DMS) Materials, R. Saravanan   Materials Research Forum LLC
Materials Research Foundations **35** (2018)   doi: http://dx.doi.org/10.21741/9781945291777

A perfect ferromagnetic system of $Ge_{1-x}Mn_x$ is the one with the right stoichiometric composition of Mn-poor Ge_8Mn_{11} and Mn. Anything that escalates the right percentage of Mn in the system of $Ge_{1-x}Mn_x$ will make it to exhibit different magnetic order. This enables us to predict the right amount of dilution of the dopant Mn in the host lattice system of Ge in order to make $Ge_{1-x}Mn_x$ to behave like a FM system. From our experimental observations, it is substantiated that, the dilution of the dopant Mn in the host matrix of Ge should not exceed 6% ($x = 0.06$) in order to achieve a material having ferromagnetic order.

6.2.1 Magnetism in $Ge_{1-x}Mn_x$: Correlation to the structure

From the present work, it is concluded that, in order to have the system behave as a ferromagnet, the dilution of Mn in the host Ge lattice should not exceed 6% ($x = 0.06$). This is because; the dilution of the dopant Mn more than the limit of 6% introduces paramagnetism in the system thus inducing the system to behave a soft ferromagnet. This can be established from the observation of the lattice disorder introduced in the system due to the presence of Mn, Mn-rich and Mn-poor metallic domains. The change in the lattice parameters for various dopant concentrations (Table 3.7) shows that when the dopant concentration is 4% (*i.e.*, $x = 0.04$), the host lattice Ge is compressed and hence, the lattice constant decreases (for $x = 0.04$, a $= 5.6469$ Å; for undoped Ge, (a $=$ 5.6558 Å). When the dopant concentration is 6% (*i.e.*, $x = 0.06$) the lattice is enlarged (a $= 5.6676$ Å) and then it decreases slightly (a $= 5.6645$ Å) when the concentration of the dopant is 10% (*i.e.*, $x = 0.10$). The expansion of the lattice due to the intermetallic domains improves the FM behavior and this confirms the present experimental observation in $Ge_{1-x}Mn_x$ that, the system behaves as a perfect ferromagnet when the dopant concentration is 6% (*i.e.*, $x = 0.06$). The B_{iso} values (Table 3.7) show that, the static disorder in the chosen system increases as x increases up to 0.04, and then decreases at $x = 0.06$ and then increases thereafter. This confirms the quantitative presence of disorder due to the presence of the intermetallic domains. The lower value of B_{iso} for $Ge_{0.94}Mn_{0.06}$ indicates that the lattice disorder is small. Hence, it can be concluded that the dilution of 6% of Mn in the Ge host matrix is the best possible dilution content for FM candidate among the Ge based DMS systems.

6.2.2 Magnetism in $Ge_{1-x}Mn_x$: Correlation to the charge density

The dopant atom Mn has atomic number and valency smaller than those of Ge but has a larger covalent radius. This makes the charge density around the host atom Ge low as the Ge atoms are replaced in the host matrix. But, due to the presence of intermetallic domains Ge_3Mn_5 and Ge_8Mn_{11}, the metallic nature in the compound increases because of the doping of Mn. This results in the increase in the interatomic charge density. When the Mn doping concentration is 4%, the number of intermetallic domains is comparatively small. Hence, the dopant atom (Mn) replacing the host Ge atom at the host lattice site decreases the charges present there. Also, there is a decrease in the mid bond charge density. With the increase in the concentration of the dopant, the interatomic region becomes clouded with localized charges due to the formation of the intermetallic domains and the Andersons mixing effect (Weng et al., 2005) between the d orbital and the delocalized s orbital of the host. Hence, the accumulation of charges in the mid bond region is less when the dopant concentration is 4% (*i.e.*, x = 0.04) but more when the dopant concentration is 6% (*i.e.*, x = 0.06). The formation of the intermetallic domains leads to the half metal like behavior of $Ge_{1-x}Mn_x$.

When x = 0.10, the lack of interactomic charges confirms that, the dilution of Mn in the host lattice Ge is minimum and the critical concentration of obtaining room temperature ferromagnetism in $Ge_{1-x}Mn_x$ is around x = 0.06. These results are supported by the theoretical predictions made by Weng et al., (2005) in which they have predicted the critical limit to be x = 0.0625. Also, when the dopant concentration is increased above this value (*i.e.*, 6%) the Mn-rich phases dominate and introduces disorder in the system. From the charge density mapping of $Ge_{1-x}Mn_x$, it is clearly observed that, the charge density at the bond critical point (3, −1) confirms the presence of the localized charges in the mid bond region (Table 5.2). The hybridization of the orbitals and the presence of the intermetallic domains (Ge_3Mn_5 and Ge_8Mn_{11}) cause the charge density to be 0.340 $e/Å^3$ when the dopant concentration is x = 0.06. But, the charge densities are less when x = 0.04 (0.322 $e/Å^3$) and x = 0.10 (0.317 $e/Å^3$). This is correlated with the expansion and the compression of the host lattice which allows the enhancement of the FM order in $Ge_{1-x}Mn_x$. Also, a unique ordering of spatial charges appeared in the system of $Ge_{1-x}Mn_x$ which enables it to behave magnetically as describe above.

6.3 Magnetic properties of $Ge_{1-x}V_x$

The measurements of the magnetic hysteresis of the melt grown samples of $Ge_{1-x}V_x$ (x = 0.03, 0.06, 0.09) were carried out using a vibrating sample magnetometer (VSM) (Lake Shore Make, Model 7410) at Sophisticated Analytical Instrument Facility (SAIF), Indian Institute of Technology, Chennai, Tamil Nadu, India, up to a magnetic field of 20kG at a constant temperature of 295K. The values of the parameters obtained from the hysteresis measurements are presented in table 6.2 and the hysteresis graphs of the samples are plotted and shown in figure 6.2 It can be clearly observed that, the sample having the least dopant concentration (*i.e.*, x = 0.03) exhibits a diamagnetic behavior with the value of magnetization being 0.24979 emu/g. When the dopant concentrations are 6% (*i.e.*, x =

0.06) and 9% (*i.e.*, x = 0.09), the samples show antiferromagnetic (AFM) behavior with the magnetization values of 0.04479 emu/g and 0.15266 emu/g respectively.

Table 6.2 Magnetic parameters of $Ge_{1-x}V_x$.

System	Coercivity (G)	Magnetization (emu/g)	Retentivity (emu/g) $(x10^{-3})$
$Ge_{0.97}V_{0.03}$	1380.4	0.24979	6.6162
$Ge_{0.94}V_{0.06}$	8181.5	0.04479	8.9043
$Ge_{0.91}V_{0.09}$	9119.3	0.15266	8.6475

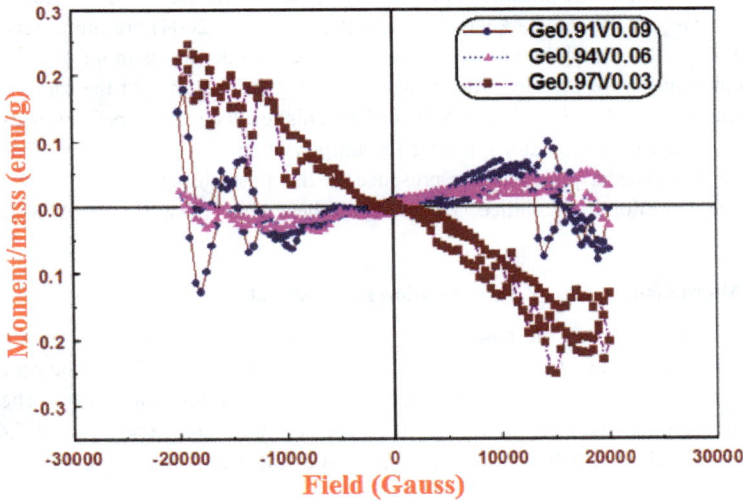

Figure 6.2 Variation of magnetic hysteresis with magnetic field in $Ge_{0.97}V_{0.03}$, $Ge_{0.94}V_{0.06}$ and $Ge_{0.91}V_{0.09}$.

When the host Ge is doped with tatoms which are paramagnetic 3d-transition metals, the interaction between the paramagnetic solute atoms and the host atoms may result in a possible formation of spin-density wave (Overhauser, 1960). Also, the dynamical readjustment of the conduction electrons always screens the interactions between the spins of the solute atoms (Hart, 1957). The resulting interaction is restricted to the nearest neighbors of the atom and the long-range part of it oscillates in sign and rapidly diminishes in magnitude with distance (Yosida, 1957). This results in a uniform flat charge between the atoms. The long-range order will result in the rearrangement of atoms which is a small distortion in the host matrix of Ge. This causes a strain in the lattice that makes the lattice to be stretched out.

Antiferromagnetic behavior of the samples is observed when $x = 0.06$ and $x = 0.09$ and diamagnetic nature is observed when the concentration of the dopant is $x = 0.03$.

It can be noted that the observed magnetic hysteresis curves of the prepared samples (Figure 6.2) show some undulations/jumps in the loops. This series of random, small jumps in the magnetization are called Barkhausen jumps (Gracía et al., 2004). These jumps may be due to the presence of crystallographic defects such as dislocations. In the sample of $Ge_{0.91}V_{0.09}$, the Barkhausen jumps (Gracía et al., 2004) are more which show that the dislocations may be more. When the dislocation density is more, the grains will start dominating to behave in a different way of magnetic order. At the same time, the compound tries to align its spins in AFM order. This results in a competition between the host system and the dislocations in which a compensation effect takes place. One reason for these Barkhausen jumps is the presence of the possible spin-density wave which causes the distortion in the lattice. However, all these effects result in the AFM behavior of $Ge_{1-x}V_x$.

6.3.1 Magnetism in $Ge_{1-x}V_x$: Correlation to the structure

The interaction between the host Ge and the paramagnetic solute V causes a possible static spin-density wave to arise. This may result in a small distortion in the host matrix of Ge and the strain in the lattice makes it to be stretched out (Table 3.8). The lattice parameters for the samples having $x = 0.06$ and $x = 0.09$ are 5.6558 Å and 5.6585 Å respectively and these values are higher than that of the sample having $x = 0.03$ (5.6543 Å).

6.3.2 Magnetism in $Ge_{1-x}V_x$: Correlation to the charge density

The long-range antiferromagnetic (AFM) order in $Ge_{1-x}V_x$ causes the charge density of the conduction electron gas to remain uniform in space. When the dopant (V) concentration is 3% ($x = 0.03$), the resultant one-dimensional charge density profile is a perfect saddle (Figure 5.7). When the dopant concentration is 6% ($x = 0.06$) and 9% ($x = 0.09$), the long-range AFM order makes the charge density uniform throughout the region. Hence, the one-dimensional charge density profile is uplifted. This results in the AFM nature in $Ge_{1-x}V_x$. The flat uniform charge density is the signature of the AFM behavior and it is clearly shown in the hysteresis behavior of the materials (Figure 6.2) when $x = 0.06$ and 0.09. But, when $x = 0.03$ the system remains diamagnetic in nature.

6.4 Magnetic properties of $Ge_{1-x}Co_x$

The magnetic measurements of the melt grown samples of $Ge_{1-x}Co_x$ (x = 0.03, 0.06, 0.09) were carried out using the vibrating sample magnetometer of Lakeshore Make, Model 7410 at Sophisticated Analytical Instrument Facility (SAIF), Indian Institute of Technology, Chennai, Tamil Nadu, India, up to a magnetic field of 20kG at a constant temperature of 295K. The recorded magnetic parameters are tabulated in table 6.3 and the observed magnetic behavior of $Ge_{1-x}Co_x$ is shown in figure 6.3.

The hysteresis graph of the sample having dopant concentration of 3% (x = 0.03) exhibits a diamagnetic behavior. $Ge_{1-x}Co_x$ having the dopant concentrations x = 0.06 and x = 0.09 exhibit antiferromagnetic behavior. The mechanism of the antiferromagnetic behavior can be explained in the following way: The introduction of the transition metal (TM) impurity (Co) in the host germanium matrix causes a magnetic moment localized at the TM site with the host Ge atoms as neighbors. Any possible hybridization between the host atoms and the orbitals of the TM atoms leads to the reduction in the local magnetic moment of the TM atom. An electronic charge transfer also takes place from the host atom to the TM atom and it occurs between the s orbital to d orbital of the dopant TM atom. The magnetic moment is mainly contributed by the d orbital of Co atom and the magnetic interaction of it with the nearest Ge atoms (Neha Kapila et al., 2011). In a similar work, when the dopant Co is introduced in the Si host lattice, the electrons in the donor atoms jump through states of the acceptor levels in the overlapping area (Collver, 1977) while the anti-bonding spin down states lie closer to the valence band. As a result, there is a competition between the super exchange and double exchange mechanisms in which the super exchange mechanism dominates and this leads to the AFM state (Kaczkowski et al., 2009). The same mechanism can be attributed to the mechanism behind the AFM behavior of $Ge_{1-x}Co_x$ investigated in this research work.

Table 6.3 Magnetic parameters of $Ge_{1-x}Co_x$.

System	Coercivity (G)	Magnetization (emu/g)	Retentivity (emu/g) $(x10^{-3})$
$Ge_{0.97}Co_{0.03}$	7.5349	0.3284	0.0101
$Ge_{0.94}Co_{0.06}$	3884.6	0.0664	6.0271
$Ge_{0.91}Co_{0.09}$	349.80	0.0633	3.7564

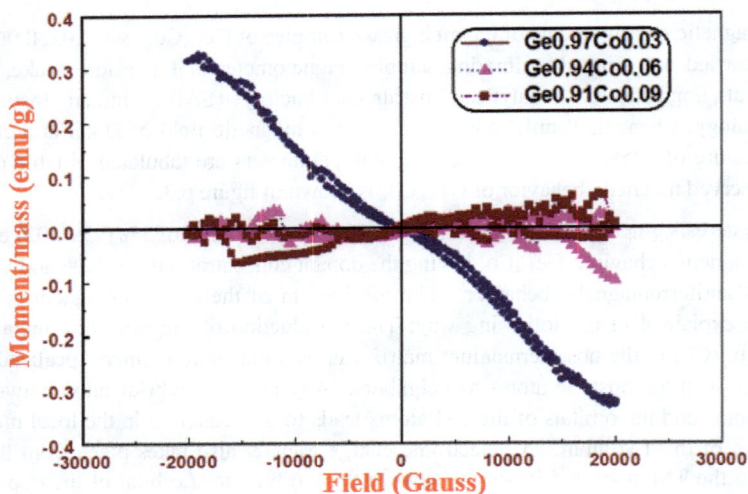

Figure 6.3 Variation of magnetic hysteresis with magnetic field in $Ge_{0.97}Co_{0.03}$, $Ge_{0.94}Co_{0.06}$ and $Ge_{0.91}Co_{0.09}$.

The formation of the GeCo cluster also contributes to the AFM hybridization happening between the orbitals of the host Ge atom and the dopant Co atom. Even though the individual magnetic moments of the dopant Co atom can lead to a possible ferromagnetic state in $Ge_{1-x}Co_x$, the interaction between the ferromagnetic Co atom and the nearest diamagnetic Ge atoms makes the antiferromagnetic behavior to dominate. When the dopant concentrations of the system are x = 0.06 and x = 0.09, the antiferromagnetic behavior has contributions from the additional phase of monoclinic GeCo which exist in $Ge_{1-x}Co_x$ as GeCo clusters.

Even though the system is understandably antiferromangetic, the magnetic hysteresis has oscillations that are termed as Barkhausen jumps (Gracía et al., 2004) which are caused because of formation of domains, irregularities, temperature fluctuations, existence of intrinsic homogeneities in the form of magnetic clusters and anisotropy. The addition of GeCo clusters in $Ge_{1-x}Co_x$ and their influence are also the cause for the observed Barkhausen jumps (Gracía et al., 2004) in the hysteresis graphs of the samples of $Ge_{1-x}Co_x$.

Dilute Magnetic Semiconducting (DMS) Materials, R. Saravanan   Materials Research Forum LLC
Materials Research Foundations 35 (2018)   doi: http://dx.doi.org/10.21741/9781945291777

6.4.1 Magnetism in $Ge_{1-x}Co_x$: Correlation to the structure

The individual magnetic moments of the dopant Co atom makes the system a ferromagnetic one but the interaction between the Co (ferromagnetic) and Ge (diamagnetic) atoms makes the system to behave as antiferromagnetic one. Thus, when the dopant concentrations of the system are $x = 0.06$ and $x = 0.09$, the antiferromagnetic behavior has contributions from the additional phase of monoclinic GeCo which exist in the chosen system as GeCo clusters. During the evaluation of the charge density, the contribution of GeCo clusters can be effectively deconvoluted from the host but the traces will still remain. Since the GeCo clusters are antiferromagnetic (Neha Kapila et al., 2011), the deconvolution will not effectively filter out the contributions of GeCo clusters and the traces remain and will influence the average electronic structure.

6.4.2 Magnetism in $Ge_{1-x}Co_x$: Correlation to the charge density

The sample of $Ge_{1-x}Co_x$ exhibits diamagnetic behavior when the concentration of the dopant is 3% ($x = 0.03$) and it becomes antiferromagnetic when the concentration of the dopant is 6% ($x = 0.06$) and 9% ($x = 0.09$). This behavior is clearly seen from the rearrangement of charges in the lattice when the concentration of the dopant is varied.

When $x = 0.03$, $Ge_{1-x}Co_x$ clearly mimics the pure system and has closed shell interaction resulting in diamagnetic behavior. When the dopant concentrations are 6% ($x = 0.06$) and 9% ($x = 0.09$), an electronic charge transfer occurs between the host Ge atom and the dopant Co atom which triggers a shared shell interaction causing the system to behave antiferromangetic (Kaczkowski et al., 2009). The GeCo clusters are formed in the process of the preparation of $Ge_{1-x}Co_x$ and this also contributes to the AFM hybridization happening between the orbitals of the host Ge atom and the dopant Co atom.

The evidence of AFM behavior can also be seen in the charge density maps and can be quantified. In our work, the retentivity of the diamagnetic material is low as is evidenced from table 6.3. In $Ge_{0.97}Co_{0.03}$, the dopant Co is so dilute and hence behaves as a diamagnetic system. However, as the dopant concentration increases, the monoclinic GeCo cluster influences the AFM behavior in the system, which is evidenced (Table 5.6) from the decrease in the charge density at the valence region when x goes from 6% ($x = 0.06$) to 9% ($x = 0.09$). This is believed to be triggered by charge transfer in the shared interaction by Co atom in the host lattice.

6.5 Magnetic properties of $Si_{1-x}Mn_x$

The magnetic measurements of the ball milled samples of $Si_{0.98}Mn_{0.02}$ (100h and 200h) were carried out using the vibrating sample magnetometer (Lake Shore Make, Model

7407), at the Central Instrumentation Facility (CIF), Pondicherry University, Pondicherry, India, up to an applied magnetic field of 10kG. The saturation magnetization values and other parameters are tabulated in table 6.4. The magnetic behavior of $Si_{0.98}Mn_{0.02}$ (100h) and $Si_{0.98}Mn_{0.02}$ (200h) are shown in figure 6.4. Magnetism in Mn-doped Si is due to the hole-mediated ferromagnetic exchange interaction and the presence of an internal magnetic field caused by the spins of the dopant atom Mn in the host lattice of Si (Lan Anh et al., 2009). The ferromagnetic behavior of the sample at room temperature proves the presence of the dopant atoms of Mn in the host lattice site of Si.

Table 6.4 Magnetic parameters of $Si_{1-x}Mn_x$.

System	Coercivity (G)	Magnetization (emu/g)	Retentivity (emu/g)
$Si_{0.98}Mn_{0.02}$(100h)	167.55	8.2012	1.4469
$Si_{0.98}Mn_{0.02}$(200h)	218.13	7.4655	1.7127

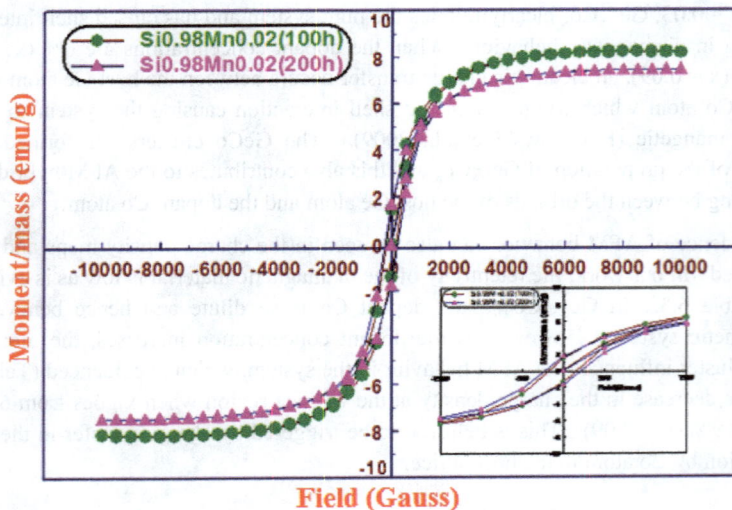

Figure 6.4 Variation of magnetic hysteresis with magnetic field in $Si_{0.98}Mn_{0.02}$(100h) and $Si_{0.98}Mn_{0.02}$(200h).

Dilute Magnetic Semiconducting (DMS) Materials, R. Saravanan   Materials Research Forum LLC
Materials Research Foundations 35 (2018)   doi: http://dx.doi.org/10.21741/9781945291777

The milling time of the samples clearly affects the magnetic properties of the samples. The change in the magnetic properties of the samples with respect to the milling time is visualized in figure 6.4. The increase in the coercivity and the decrease in the saturation magnetization with the increase in the milling time can be clearly observed. This result can be attributed to the reduction of grain size, increased internal strain and the formation of defects introduced during the milling process (Mutlu Can et al., 2010; Chen et al., 1996). Also, the variation in the coercivity is related to the evolution of magnetic domains with particle size. The saturation magnetization of ball milled $Si_{1-x}Mn_x$ decreases with decreasing size and is proportional to the specific surface area of the particles (Chen et al., 1996). The presence of the Mn^{2+} ions has localized magnetic moments and this induces the charge ordering state. This is reflected on the magnetic behavior of $Si_{1-x}Mn_x$. The doping of Mn^{2+} ions in the host lattice of Si^{4+} results in weak ferromagnetic spin correlation and this reduces the degree of magnetization. The magnetic state of the samples with higher concentration of the dopant transition metal Mn is attributed to the formation of cluster-glass state of spins. The spin glass state is attributed to the variation of the ratio of Mn^{3+}/Mn^{4+} and the surface effect due to the decrease in the particle size caused by the increase in the milling time.

6.5.1 Magnetism in $Si_{1-x}Mn_x$: Correlation to the structure

The increase in the cell dimension is clearly noted for the ball milled samples (Table 3.11) which confirms the incorporation of Mn in the host lattice of Si. Also, the additional phase of SiMn has been confirmed from the observed X-ray diffraction profile of the sample. These SiMn clusters are the reason behind the ferromagnetic behavior of the system of Si:Mn. The decrease in the particle size due to the increase in the milling time is also observed (Table 3.11). As the particle size of the system is decreased, the ferromagnetic nature of the sample is decreased and it becomes a soft ferromagnet.

6.5.2 Magnetism in $Si_{1-x}Mn_x$: Correlation to the charge density

As the Mn^{2+} ions are doped in the host system of Si, there is an increase in the hole density and this reduces the midbond charge density. The accumulation of the charges from the intermediate bonding region can increase the resistivity of the system that forms the basis of the hole-mediated ferromagnetic exchange interactions in Mn-doped Si (Lan Anh et al., 2009). When the milling time is increased, the generation of the secondary phase, SiMn is enhanced. Hence, depletion of charges takes place in the bonding region thereby forming the closed shell interaction which makes the system to enter into a near insulator phase due to the increase in the resistivity of the system. Hence, a perfect ferromagnetic state is attained in $Si_{0.98}Mn_{0.02}$ when milled 100 h and a soft ferromagnetic state when ball milled 200 h.

6.6 Magnetic properties of $Si_{1-x}Ni_x$

It is mandatory for a semiconductor to exhibit room temperature ferromagnetic (RTFM) property so as to qualify itself as a DMS material and hence, the magnetic properties of the samples are investigated in this work. The magnetic measurements of the prepared DMS samples of $Si_{1-x}Ni_x$ were carried out using vibrating sample magnetometer (VSM) (Lake Shore Make, Model 7410), at Sophisticated Analytical Instrument Facility (SAIF), Indian Institute of Technology Madras, Tamil Nadu, India, up to a magnetic field of 16 KG at a constant temperature of 295 K. The magnetic behavior for the prepared samples is shown in figure 6.5(a). The magnetic behavior of the samples of $Si_{1-x}Ni_x$ within the range of 2000G is shown in figure 6.5(b). The observed magnetic parameters of ball milled $Si_{1-x}Ni_x$ are tabulated in table 6.5.

Well pronounced ferromagnetic property is exhibited by all of the prepared samples of $Si_{1-x}Ni_x$. The samples with the dopant concentrations of x = 0.03 and 0.06 behave ferromagnetically in the same way and this is evidenced by the merging hysteresis curves of the two samples (Figure 6.5(a)). It is understood that when x = 0.03, the influence of micro clusters of secondary Ni phase is small. The ferromagnetic property becomes well pronounced as the dopant concentration is increased to 12% (Table 6.5). It is established here that, the diamagnetic pure Si makes a transition to FM as the dopant Ni is introduced in the system and the transition is found to be gradual rather than a sudden one confirming its dependence on the dopant concentration of Ni.

Table 6.5 Magnetic parameters of $Si_{1-x}Ni_x$.

System	Coercivity (G)	Magnetization (emu/g)	Retentivity (emu/g)
$Si_{0.97}Ni_{0.03}$	129.91	83.337	8.9455
$Si_{0.94}Ni_{0.06}$	109.30	83.725	6.5441
$Si_{0.91}Ni_{0.09}$	110.16	131.38	10.264
$Si_{0.88}Ni_{0.12}$	106.84	251.85	19.570

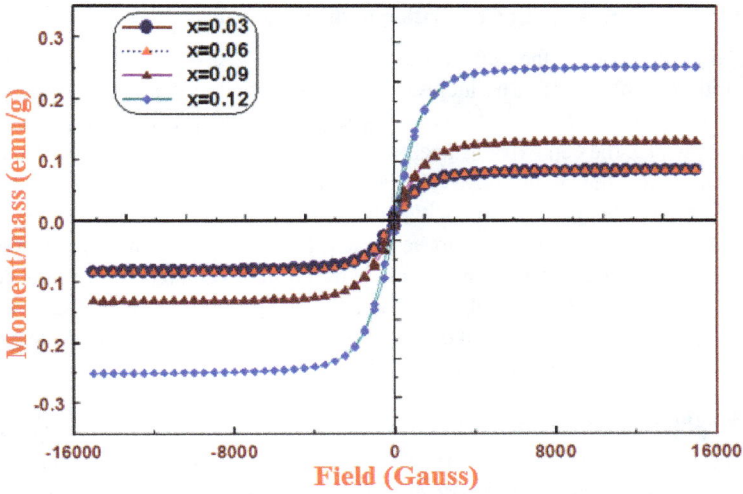

Figure 6.5 (a) Variation of magnetic hysteresis with magnetic field in $Si_{1-x}Ni_x$ (x = 0.03, 0.06, 0.09, 0.12).

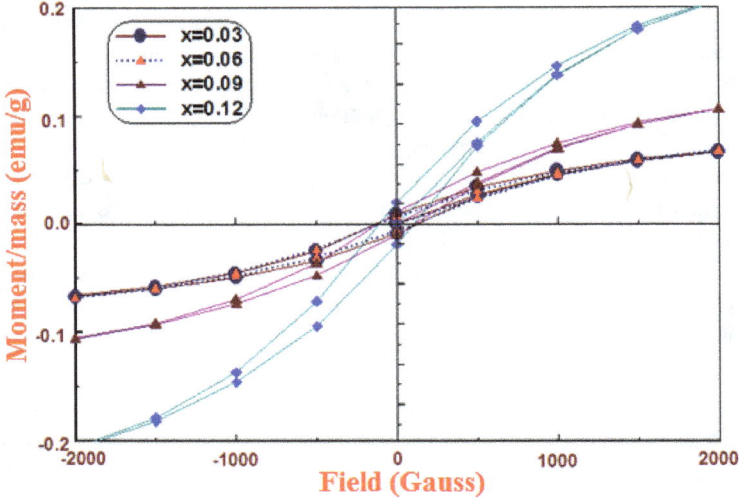

Figure 6.5(b) Variation of magnetic hysteresis with magnetic field in $Si_{1-x}Ni_x$ (x = 0.03, 0.06, 0.09, 0.12) (Within the range of 2000G).

6.6.1 Magnetism in $Si_{1-x}Ni_x$: Correlation to the structure

In $Si_{1-x}Ni_x$, the value of the cell dimension increases with the increase in the dopant concentration (Table 3.12). The inclusion of the bigger Ni (135pm (Haynes, 2014)) ion into the lattice site of host Si (110pm (Haynes, 2014)) matrix introduces strain in the system and hence, the cell expands itself to accommodate this forceful inclusion of Ni in the regular lattice site of the host. The dopant Ni prefers to be in the substitutional position instead of being in the interstitial sites. Also, the additional phase in the form of microclusters of Ni metal is found to be present along with Si:Ni in the prepared DMS system. It is also found that, as the dopant concentration is increased, the concentration of the microclusters of Ni metal also increases thereby paving way for the semiconductor to metal (S-M) transition to take place which in turn causes room temperature ferromagnetism in the ball milled $Si_{1-x}Ni_x$ DMS system.

6.6.2 Magnetism in $Si_{1-x}Ni_x$: Correlation to the charge density

As expected, the prepared systems $Si_{1-x}Ni_x$ show room temperature ferromagnetic (RTFM) property which is necessary for a DMS material. The hysteresis curves at x = 0.03 and 0.06 merge (Figure 6.5) and enact the same behavior. It is clear from the curves that, the system makes a transition from diamagnetic to ferromagnetic and RTFM is well established even at lowest Ni solubility of 3% where influence of micro clusters of secondary Ni phase is minimum. The ferromagnetic property becomes well pronounced as x increases up to 0.12. The magnetization measurements of the samples (Table 6.5) reveal that, the saturation magnetization for x = 0.03 and x = 0.06 are same whereas it increases threefold when the dopant concentration goes to x = 0.12. It should be recalled here that, the contribution of secondary phase of Ni micro clusters cannot be ruled out. However, the charge density reveals the fact that the system goes from semiconductor to metallic state and it is a result of accumulation of charges at the valence region as seen in the three dimensional (Figure 5.15(a) to (e)), two dimensional (Figure 5.16(a) to 6(e)) and one dimensional (Figure 5.17) charge density mapping. The excess charges accumulating at the valence region will not be spin compensated and thus goes to the ferromagnetic state.

When the system makes a transition from p to n type semiconductor with the increase in the dopant concentration, the impurity levels generated by the dopant Ni are not allowed to contribute to the deep level. Hence, the contribution of Ni to the bound cores is hindered thus allowing to be accommodated only along the valence region. This increases the charge density at the bond critical point. This accumulation of charges have uncompensated spins and contributes to the ferromagnetic property of the system.

6.7   Conclusion

The magnetic nature of the prepared samples is analyzed using the measurements recorded using vibrating sample magnetometer. The results are presented here.

6.7.1   Magnetism in $Ge_{1-x}Mn_x$

i.   When $x = 0.04$, antiferromagnetic behavior is exhibited by the sample.

ii.   Perfect ferromagnetic behavior is observed for the sample with $x = 0.06$.

iii.   The sample is a soft ferromagnet when $x = 0.10$.

6.7.2   Magnetism in $Ge_{1-x}V_x$

i.   When $x = 0.03$, the sample exhibits diamagnetic behavior.

ii.   Antiferromagnetic behavior is observed in the system when $x = 0.06$ and 0.09.

iii.   The presence of the static spin-density wave is attributed to the antiferromagnetic behavior of the system.

6.7.3   Magnetism in $Ge_{1-x}Co_x$

i.   When $x = 0.03$, diamagnetic behavior is observed in $Ge_{1-x}Co_x$.

ii.   When the concentration of the dopant is increased to $x = 0.06$ and 0.09, the system behaves as an antiferromagnet.

6.7.4   Magnetism in $Si_{0.98}Mn_{0.02}$

i.   Perfect room temperature ferromagnetism is observed in $Si_{0.98}Mn_{0.02}$ milled for 100h.

ii. $Si_{0.98}Mn_{0.02}$ behaves as a soft ferromagnet when milled for 200h.

iii.   The decrease in size, increase in internal strain, hole-mediated ferromagnetic interactions are reason for the mentioned magnetic behavior of the ball milled samples.

6.7.5   Magnetism in $Si_{1-x}Ni_x$

i.   Room temperature ferromagnetism is observed in all the prepared DMS samples of $Si_{1-x}Ni_x$.

ii.   The DMS samples having the dopant concentrations of $x = 0.03$ and 0.06 has similar ferromagnetic behavior.

iii.   The ferromagnetic property is well pronounced when the dopant concentration is x $= 0.12$.

iv.   Accumulation of charges is observed with the increase in dopant concentration and have uncompensated spins leading to the observed ferromagnetic nature of the prepared system.

v.   The S-M transition that depends on the dopant concentration and the p-n type transition are responsible for the observed room temperature ferromagnetic nature of the prepared DMS system of $Si_{1-x}Ni_x$.

References

[1]   Beigger E., Stäheli L., Fonin M., Rüdiger U., Journal of Applied Physics 101, 103912 (2007). https://doi.org/10.1063/1.2718276

[2]   Chen J.P., Sorensen C.M., Klabunde K.J., Hadjipanayis G.C., Devlin E., Kostikas A., Phys. Rev. B., 54, 9288 (1996). https://doi.org/10.1103/PhysRevB.54.9288

[3]   Collver M.M., Solid State Comm., 23, 333 (1977). https://doi.org/10.1016/0038-1098(77)91340-0

[4]   Gracía Calderón R., Gómez Sal J.C., Iglesias J.R., J. Mag. Mag. Mat., 701, 272 (2004).

[5]   Hart E.W., Phys. Rev., 106, 467 (1957). https://doi.org/10.1103/PhysRev.106.467

[6]   Haynes W.M., CRC Handbook of Chemistry and Physics, CRC Press/Taylor and Francis, Boca Raton, FL, 95th Edition, 2014.

[7]   Kaczkowski J., Jezierski A., Acta Physica Polonica A., 115 No. 1, 275 (2009). https://doi.org/10.12693/APhysPolA.115.275

[8]   Lan Anh T.T., Yu S.S., Ihm Y.E., Kim D.J., Kim H.J., Hong S.K., Kim C.S., Physica B 404, 1686 (2009). https://doi.org/10.1016/j.physb.2009.02.001

[9]   Mutlu Can M., Ozcan S., Ceylan A., Firat T., Mater. Sci. Eng. B., 172, 72 (2010). https://doi.org/10.1016/j.mseb.2010.04.019

[10]   Neha Kapila., Jindal V.K., Hitesh Sharma, Physica B, 406, 4612 (2011). https://doi.org/10.1016/j.physb.2011.09.038

Dilute Magnetic Semiconducting (DMS) Materials, R. Saravanan   Materials Research Forum LLC
Materials Research Foundations **35** (2018)   doi: http://dx.doi.org/10.21741/9781945291777

[11]   Overhauser A.W., J. Phys. Chem. Solids., 13, 71 (1960).
        https://doi.org/10.1016/0022-3697(60)90128-1

[12]   Weng H., Dong J., Physical Review B 71, 035201 (2005).
        https://doi.org/10.1103/PhysRevB.71.035201

[13]   Yosida K., Phys. Rev., 106, 893 (1957). https://doi.org/10.1103/PhysRev.106.893

Dilute Magnetic Semiconducting (DMS) Materials, R. Saravanan   Materials Research Forum LLC
Materials Research Foundations **35** (2018)   doi: http://dx.doi.org/10.21741/9781945291777

Chapter 7

Local Structure of Prepared DMS Materials

Abstract

Chapter 7 deals with the analysis of local structure of the prepared DMS materials using the observed powder X-ray diffraction intensities. This analysis is done using the pair distribution function of the materials. The distances of the nearest neighbors are tabulated, which are helpful in quantifying any deviation or strain in the lattice, which can possibly occur due to the process of doping.

Keywords

Local Structure, Pair Distribution Function (PDF), Observed X-ray Intensity, Strain, Nearest Neighbor

Contents

7.1   Introduction

The local structure of a material gives information on the arrangement of atoms over a scale of few interatomic distances. At these small distances, any variation in the crystal structure will make the local arrangement of atoms to appear significantly different from the average structure that is revealed by the X-ray diffraction techniques. Local structural analysis is very important in the case of disordered materials or short-range materials such as glasses, fluids and fine powders. The analysis can also be done for crystalline materials having disorder in the orientations of atoms and magnetic moments, positions of atoms *etc*. It is essential to carry out the local structural analysis in the case of diluted magnetic semiconductor materials since any inclusion of the dopant transition metal atoms in the host lattice of a semiconductor will change the local arrangement of atoms in a significant manner.

The local structural analysis of a material is done by using the pair distribution function (PDF) (Proffen et al., 1999) which reveals the structure around a specific atom in a detailed way. For the calculation of the pair distribution function, an initial structural model of the atomic arrangements is considered. Then, by optimizing the model parameters, a PDF which is in good agreement with the experimental one is obtained and then analyzed. The relative intensity in the PDF is related directly to the number of pairs of atoms. The point at which there are no peaks in the PDF is considered as the extent of the long range ordering within the material.

For the calculation and the analysis of the pair distribution function (PDF) (Proffen et al., 1999) of the prepared diluted magnetic semiconductors, the software package PDFgetX (Jeong et al., 2001) has been used where the raw data has been converted into a convenient format for the required analysis to be done within a convenient range and step size. The software PDFgui (Farrow et al., 2007) has been used for the refinement procedure in which the theoretical structure models have been fitted with the experimental PDF. This refinement procedure includes the structural parameters such as phase, lattice dimension, scale factor, linear and quadratic atomic correlation factors along with occupancy of atoms and the thermal amplitudes. The detailed analysis of the local structure of the prepared Si and Ge based DMS materials are presented in the following sections.

7.2   Analysis of the local structure of $Ge_{1-x}Mn_x$

The powder X-ray diffraction data of the melt grown samples of Ge, $Ge_{0.96}Mn_{0.04}$, $Ge_{0.94}Mn_{0.06}$ and $Ge_{0.90}Mn_{0.10}$ were used for the analysis of the pair distribution function using the software PDFgetX (Jeong et al., 2001). The theoretical model was fitted with the experimental one using the software PDFgui (Farrow et al., 2007) in a refinement procedure which includes the factors such as phase, scale factor, linear and quadratic atomic correlation factors, lattice parameters, *etc*. The refinement process was carried out within the interatomic distance of r = 30 Å with a step size of 0.02. The upper limit for the Fourier transformation of the data was considered as Q_{max} = 6.9 Å^{-1}, where $Q = (4\pi \sin\theta)/\lambda$. The observed and the calculated PDFs for the samples of Ge, $Ge_{0.96}Mn_{0.04}$, $Ge_{0.94}Mn_{0.06}$ and $Ge_{0.90}Mn_{0.10}$ are compared and the nearest neighbor distances for the samples are evaluated and tabulated in table 7.1. The comparison of the observed pair distribution function of Ge, $Ge_{0.96}Mn_{0.04}$, $Ge_{0.94}Mn_{0.06}$ and $Ge_{0.90}Mn_{0.10}$ is shown in figure 7.1. The fitted profiles of the samples are presented in figures 7.2 (a) to (d).

It is clear from the tabulated values (Table 7.1) that, the bond-lengths of the systems increase with the increase in the concentration of the dopant Mn in the host matrix of Ge. When the concentration of the dopant Mn increases in $Ge_{1-x}Mn_x$, the peak intensities of the pair distribution function of the systems decrease which substantiates the incorporation of the dopant Mn atom in the host lattice of Ge. The mismatch in the bond-lengths of the 3rd nearest neighbor is due to the existence of the Ge_3Mn_5 phase in the DMS material $Ge_{1-x}Mn_x$. The undulations in the PDFs are also attributed to the existence of the additional phase.

Table 7.1 Bond lengths obtained from PDF analysis for $Ge_{1-x}Mn_x$.

System	Ist neighbour (Å)		ΔD (%)	IInd neighbour (Å)		ΔD (%)	IIIrd neighbour (Å)		ΔD (%)
	D_{obs}	D_{cal*}		D_{obs}	D_{cal*}		D_{obs}	D_{cal*}	
Ge	2.40	2.45	2.04	4.08	4.00	2.00	5.94	5.66	4.95
$Ge_{0.96}Mn_{0.04}$	2.50	2.45	2.04	4.06	4.00	1.50	5.98	5.66	5.65
$Ge_{0.94}Mn_{0.06}$	2.54	2.45	3.67	4.22	4.00	5.50	5.98	5.66	5.65
$Ge_{0.90}Mn_{0.10}$	2.56	2.45	4.49	4.22	4.00	5.50	6.02	5.66	6.36

D_{obs}-Observed nearest neighbor distance
D_{cal*} - Calculated nearest neighbor distance
ΔD-Percentage deviation of the nearest neighbor distance
* Using theoretical model (Gretep Software) (Bochu, 2010)

Dilute Magnetic Semiconducting (DMS) Materials, R. Saravanan   Materials Research Forum LLC
Materials Research Foundations **35** (2018)   doi: http://dx.doi.org/10.21741/9781945291777

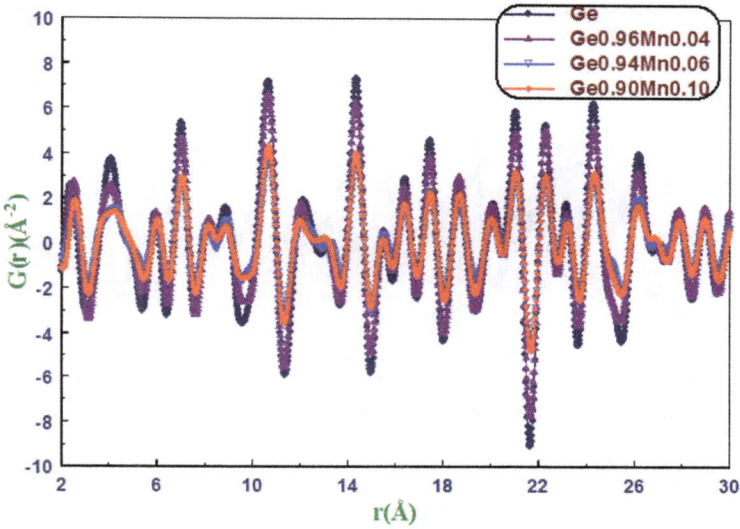

Figure 7.1 Variation of PDF peak intensity with increase in concentration of Mn in $Ge_{1-x}Mn_x$.

(a)

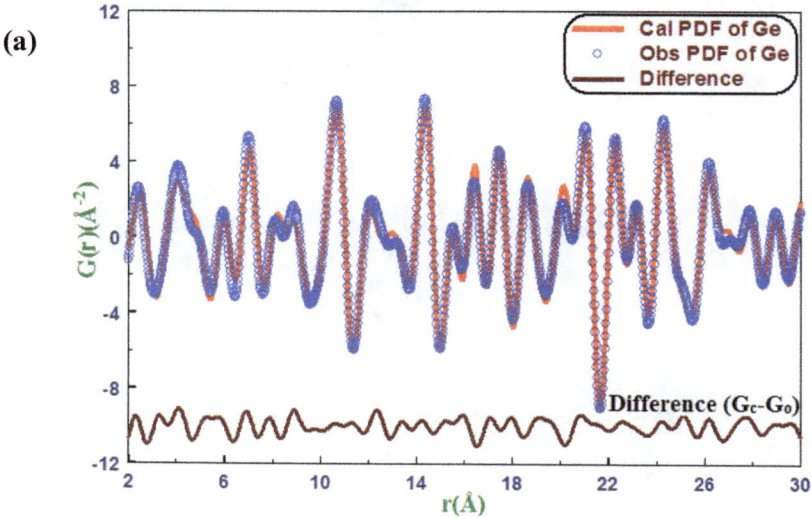

Figure 7.2(a) Fitted pair distribution function (PDF) of Ge.

161

(b)

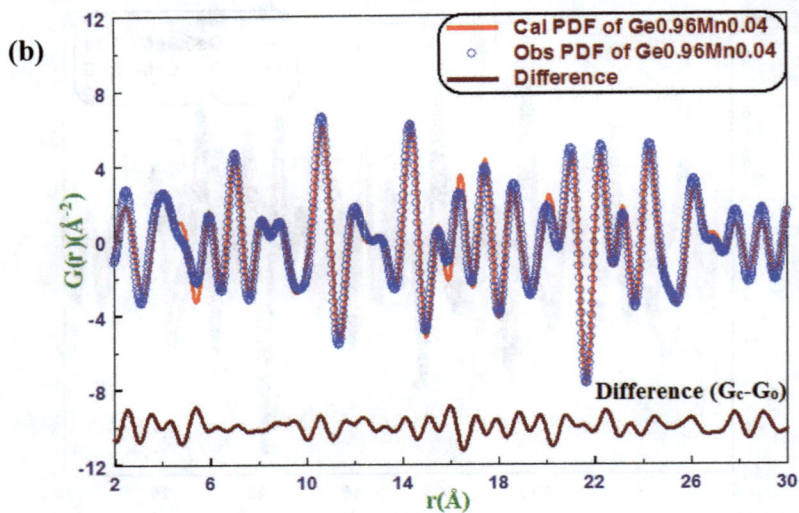

Figure 7.2(b) Fitted pair distribution function (PDF) of Ge$_{0.96}$Mn$_{0.04}$.

(c)

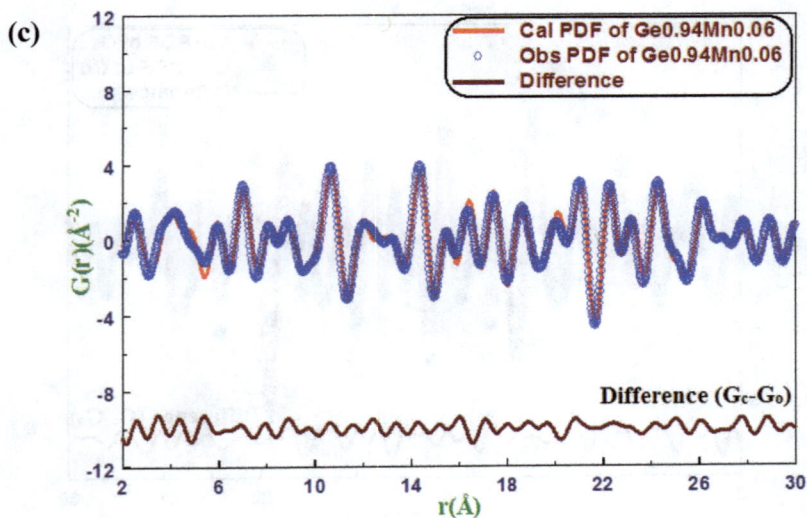

Figure 7.2(c) Fitted pair distribution function (PDF) of Ge$_{0.94}$Mn$_{0.06}$.

Dilute Magnetic Semiconducting (DMS) Materials, R. Saravanan   Materials Research Forum LLC
Materials Research Foundations **35** (2018)   doi: http://dx.doi.org/10.21741/9781945291777

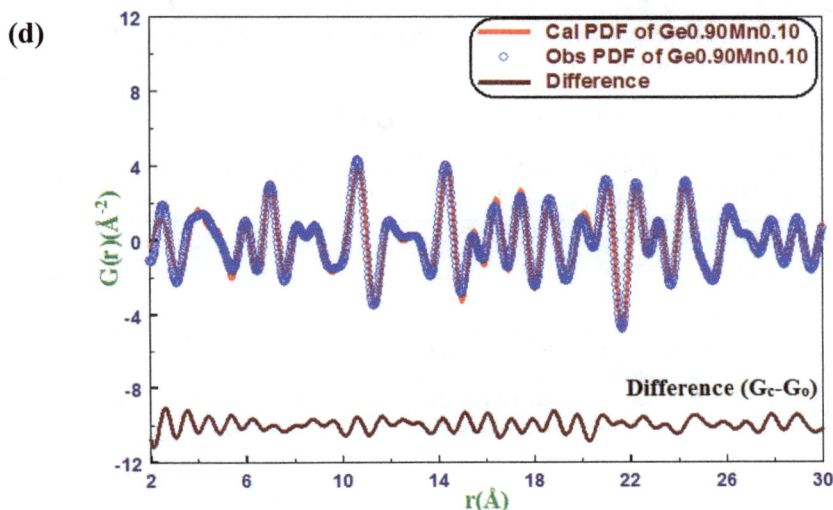

Figure 7.2(d) Fitted pair distribution function (PDF) of $Ge_{0.90}Mn_{0.10}$.

7.3   Analysis of the local structure of $Ge_{1-x}V_x$

The pair distribution function (PDF) (Proffen et al., 1999) of $Ge_{1-x}V_x$ (x = 0.03, 0.06, 0.09) has been evaluated from the experimentally observed powder X-ray diffraction data by employing the software package PDFgetX (Jeong et al., 2001). The observed and the calculated PDF were compared using PDFgui (Farrow et al., 2007). The comparison of the pair distribution functions was done in a refinement process that involves the refinement of lattice parameter, phase, scale factor, linear and quadratic atomic correlation factors along with the thermal amplitudes and occupancy of atoms.

In the present work, the refinement process has been carried out within the interatomic distance of r = 30 Å with a step size of 0.02 Å. The upper limit for the Fourier transform of the data was considered as Q_{max} = 7 Å^{-1}, where $Q = (4\pi \sin\theta)/\lambda$. Then, the calculated PDFs were compared with that of the observed ones to analyze the local structure of $Ge_{1-x}V_x$. The separation between the nearest neighbor atoms in $Ge_{0.97}V_{0.03}$, $Ge_{0.94}V_{0.06}$ and $Ge_{0.90}V_{0.10}$ are tabulated and presented in table 7.2. The observed and the calculated PDFs for the chosen samples are compared and shown in figure 7.3.

It can be clearly noted from the plotted pair distribution functions that (Figure 7.3), there are undulations below 13 Å. These undulations confirm the presence of vanadium and

Dilute Magnetic Semiconducting (DMS) Materials, R. Saravanan   Materials Research Forum LLC
Materials Research Foundations 35 (2018)   doi: http://dx.doi.org/10.21741/9781945291777

the dislocation due to the implantation of vanadium atoms in the host system of germanium. These observations are acting as evidences for the antiferromagnetic (AFM) behavior of the chosen material of $Ge_{1-x}V_x$.

Table 7.2 Bond lengths obtained from PDF analysis for $Ge_{1-x}V_x$.

System	I^{st} neighbour(Å)			IInd neighbour(Å)		
	D_{obs}	D_{cal*}	ΔD (%)	D_{obs}	D_{cal*}	ΔD (%)
$Ge_{0.97}V_{0.03}$	2.26	2.45	7.75	4.02	4.00	0.50
$Ge_{0.94}V_{0.06}$	2.30	2.45	6.12	4.04	4.00	1.00
$Ge_{0.91}V_{0.09}$	2.30	2.45	6.12	4.06	4.00	1.50

D_{obs}-Observed nearest neighbor distance
D_{cal*}-Calculated nearest neighbor distance
ΔD-percentage deviation of the nearest neighbor distance
* Using theoretical model (Gretep Software) (Bochu, 2010)

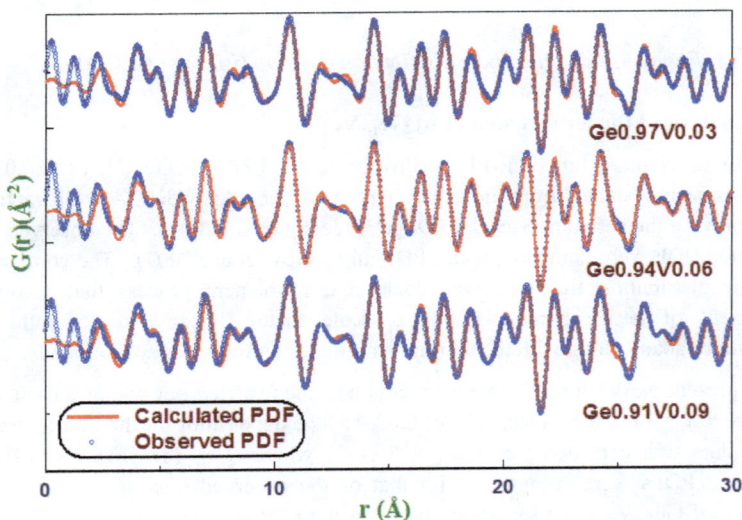

Figure 7.3 Comparison of the observed and the calculated pair distribution functions (PDFs) of $Ge_{0.97}V_{0.03}$, $Ge_{0.94}V_{0.06}$ and $Ge_{0.91}V_{0.09}$.

7.4 Analysis of the local structure of $Ge_{1-x}Co_x$

The structural changes due to doping can be extracted using the pair distribution function (Proffen et al., 1999) analysis in which the structure related information extracted from powder diffraction data can be used to study the local structure of a crystalline or amorphous material.

In the present work, the pair distribution function analysis has been done using the observed powder X-ray diffraction data of the melt grown samples of $Ge_{0.97}Co_{0.03}$, $Ge_{0.94}Co_{0.06}$ and $Ge_{0.91}Co_{0.09}$ by employing the software package PDFgetX (Jeong et al., 2001). The software PDFgui (Farrow et al., 2007) has been used to fit the theoretical structure model with the experimental one. In our refinement procedure involving the structural parameters such as lattice parameter, phase, scale factor, linear and quadratic atomic correlation factors, thermal amplitudes and occupancy of atoms, the interatomic distance for refinement was considered as r = 30 Å with a step size of 0.02 Å. The upper limit for the Fourier transform of the data was taken as $Q_{max} = 7$ Å^{-1}, where $Q = (4\pi \sin\theta)/\lambda$. The calculated PDF is compared with the observed PDF and shown in figure 7.4. The nearest neighbor distances for the samples of $Ge_{0.97}Co_{0.03}$, $Ge_{0.94}Co_{0.06}$ and $Ge_{0.91}Co_{0.09}$ are evaluated and presented in table 7.3. From the plotted PDF profiles, it can be clearly observed that, the intensity of the peaks decreases with the increase in the concentration of the dopant Co in the host lattice of Ge. This evidences the incorporation of the dopant Co atom in the host lattice of Ge. The deviations in the profile up to 8 Å ensures the fact that, GeCo clusters are formed which are so small but rather effective in deciding the magnetic behavior of $Ge_{1-x}Co_x$.

Table 7.3 Bond lengths obtained from PDF analysis for $Ge_{1-x}Co_x$.

System	Ist neighbour(Å)		ΔD (%)	IInd neighbour(Å)		ΔD (%)	IIIrd neighbour(Å)		ΔD (%)
	D_{obs}	D_{cal^*}		D_{obs}	D_{cal^*}		D_{obs}	D_{cal^*}	
$Ge_{0.97}Co_{0.03}$	2.22	2.00	11.00	4.02	4.00	0.50	5.94	5.65	5.13
$Ge_{0.94}Co_{0.06}$	2.14	2.00	7.00	2.98	3.46	13.87	3.98	4.00	0.50
$Ge_{0.91}Co_{0.09}$	2.12	2.00	6.00	2.98	3.46	13.87	3.98	4.00	0.50

D_{obs}-Observed nearest neighbor distance
D_{cal^*}-Calculated nearest neighbor distance
ΔD-percentage deviation of the nearest neighbor distance
* Using theoretical model (Gretep Software) (Bochu, 2010)

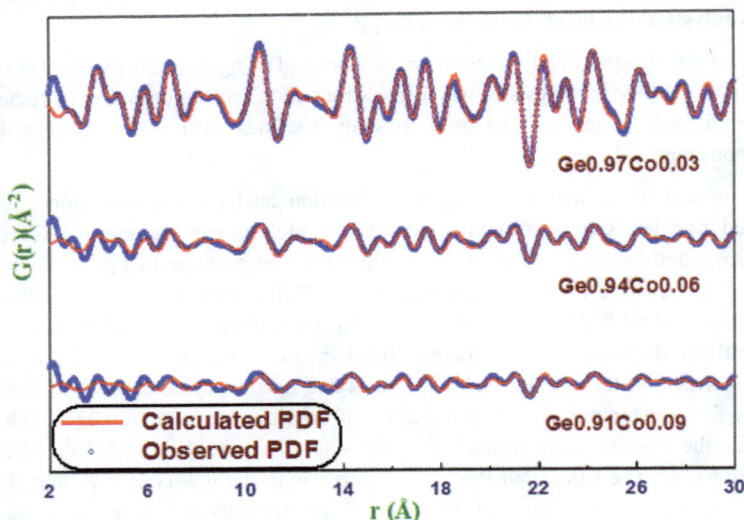

Figure 7.4 Comparison of the observed and the calculated pair distribution functions (PDFs) of $Ge_{0.97}Co_{0.03}$, $Ge_{0.94}Co_{0.06}$ and $Ge_{0.91}Co_{0.09}$.

7.5   Analysis of the local structure of $Si_{1-x}Mn_x$

The pair distribution function (Proffen et al., 1999) analysis enables us to gather information on the arrangement of atoms in poorly crystalline materials such as the ball milled samples of $Si_{1-x}Mn_x$ analyzed in this work.   The pair distribution function is generated from the sine Fourier transform of the normalized scattering function obtained from the powder X-ray diffraction data of the samples under study.   Plotting the PDF gives the probability of finding an atom at a given distance 'r' from another atom in the material under investigation.

In this work presented here, the observed powder X-ray diffraction data sets of the samples have been used for the analysis of the pair distribution function (PDF) (Proffen et al., 1999).   The observed pair distribution functions of the samples have been obtained using the software package PDFgetX (Jeong et al., 2001).   This software enables us to reduce the raw data into a convenient format for the analysis to be done with a convenient range of interatomic distance and step size.   The refinement process has been carried out within the interatomic distance of $r = 3.5$ Å to 25 Å and a step size of 0.02 Å.   The upper limit for the sine Fourier transform of the data has been considered as $Q_{max} = 7.05$ Å^{-1}, where $Q = (4\pi \sin \theta)/\lambda$. The software PDFgui (Farrow et al., 2007) has been used for fitting the PDFs. The nearest neighbor distances for the samples of Si, $Si_{0.98}Mn_{0.02}$ (100h) and $Si_{0.98}Mn_{0.02}$ (200h) are tabulated in table 7.4. The observed pair distribution functions of the ball milled samples are compared and shown in figure 7.5. The fitted pair distribution of the samples of Si and $Si_{0.98}Mn_{0.02}$ (100h) are shown in figures 7.6 (a) and (b) respectively.

It is evident from the profiles of the observed pair distribution function of the samples (Figure 7.5) that, the reduction in the intensity of the peaks of the PDF results in the estimation of number of atoms at a neighbor distance with respect to reduction of size due to the mechanical alloying process. The integration of the peaks is a source of information about the number of atoms present in a particular distance. This also gives an idea on how the reduction in the size of the grain results in the reduction in the number of atoms at a selected neighbor distance. From the observed profiles, it was calculated that with an increase in the time of milling, the number of atoms at a selected neighbor distance is reduced to 57% (100h) and then to 4% (200h) indicating that the milling of the sample affects the amplitude of the PDF profile with respect to the reduction in the size of the grain. The pair distribution function derived from the X-ray information shows that all the three systems behave the same way. This fact is shown by the matching of wriggles in the profile while the amplitude of the function decreases as the ball milling time is increased. The observed and calculated PDFs are fitted for the samples of Si and ball milled $Si_{0.98}Mn_{0.02}$ (100h) whereas the peaks of the pair distribution function of the sample of ball milled $Si_{0.98}Mn_{0.02}$ (200h) is too low to fit the profiles and hence not presented here.

Table 7.4 Bond lengths obtained from PDF analysis for $Si_{1-x}Mn_x$.

r (Si) (Å)	r ($Si_{0.98}Mn_{0.02}$(100h)) (Å)	r ($Si_{0.98}Mn_{0.02}$(200h)) (Å)
3.92	3.90	3.94
5.76	5.76	5.80
6.72	6.68	6.70
7.78	7.72	7.68
8.42	8.48	8.56
10.20	10.26	10.32

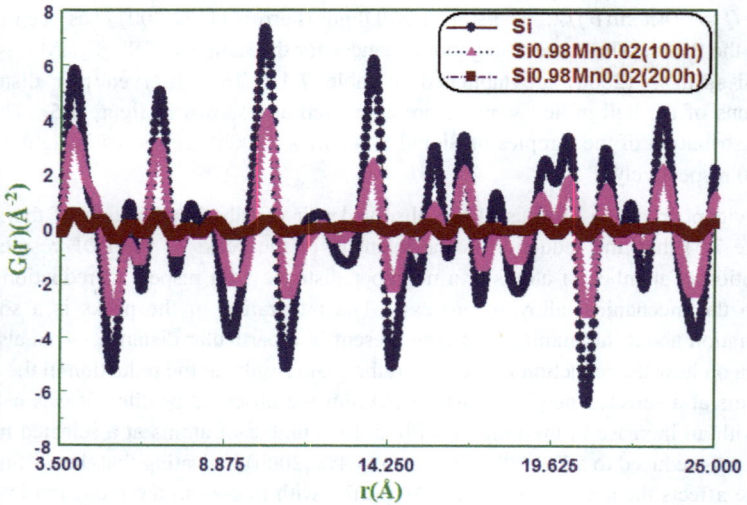

Figure 7.5 Comparison of the observed pair distribution function (PDF) of the samples Si, Si$_{0.98}$Mn$_{0.02}$ (100h) and Si$_{0.98}$Mn$_{0.02}$ (200h).

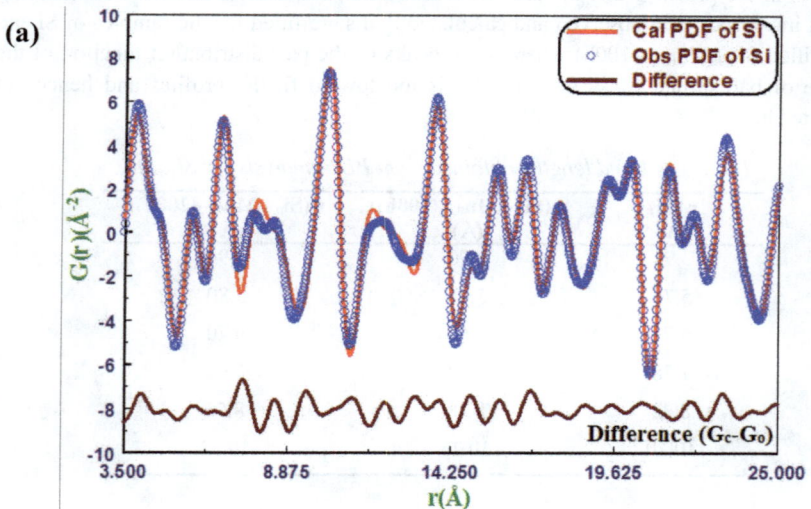

Figure 7.6(a) Fitted pair distribution function (PDF) profiles of Si

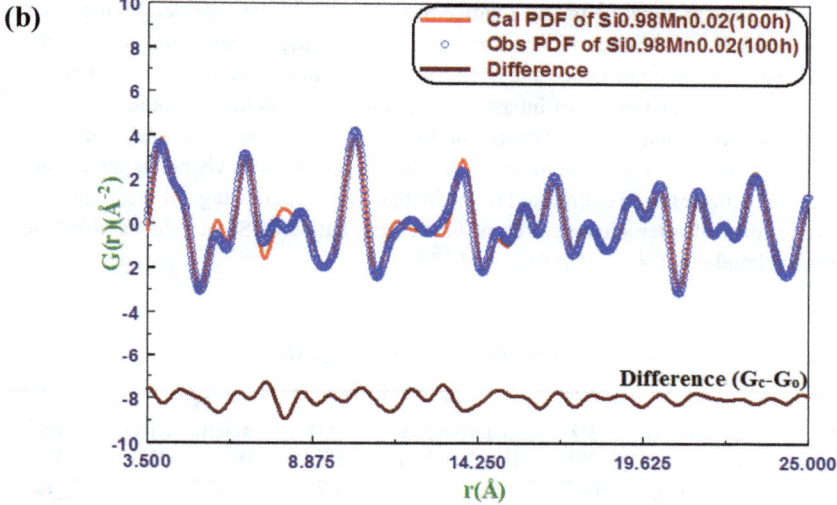

Figure 7.6(b) Fitted pair distribution function (PDF) profiles of $Si_{0.98}Mn_{0.02}$ (100h).

7.6 Analysis of the local structure of $Si_{1-x}Ni_x$

An important technique that can extract structural changes due to doping is called pair distribution function (PDF) (Proffen et al., 1999) analysis. It is very useful to study the local structure of a crystalline or amorphous material extracted from the observed powder X-ray diffraction data. Hence, it can provide information about short-range (<10nm) ordering in materials, in particular, information about bond-length distribution and lattice constants and also the differences in structural properties due to doping.

The pair distribution function analysis has been done using the observed powder X-ray diffraction data of the samples of $Si_{1-x}Ni_x$ (x = 0, 0.03, 0.06, 0.09, 0.12) employing software package PDFgetX (Jeong et al., 2001). The software package PDFgui (Farrow et al., 2007) employs a method of fitting a trial theoretical structure model with the experimental PDF by refining structural parameters such as lattice parameter, phase scale factor, linear and quadratic atomic correlation factors along with thermal amplitudes and occupancy of atoms. In our work, the refinement process has been carried out within the interatomic distance of r = 30 Å with a step size of 0.02 Å with the upper limit for the Fourier transform of the data as Q_{max} = 7Å^{-1}, where $Q = (4\pi \sin\theta)/\lambda$. The fitted observed PDFs of the samples are compared in figures 7.7 (a) to (e) and the nearest neighbor distances for the samples are evaluated and presented in table 7.5.

Materials Research Forum LLC
doi: http://dx.doi.org/10.21741/9781945291777

As the dopant Ni is included in the host lattice Si (at $x = 0.03$), the bond length increases due to the expansion of lattice in order to accommodate the forceful inclusion of the bigger Ni atom in the regular lattice site of Si. As the dopant concentration increases, the presence of the additional phase of interstitial metallic Ni microclusters comes into play which reduces bond length. The mismatch in the observed and the calculated profiles of pair distribution functions of the samples of $Si_{1-x}Ni_x$ becomes more visible as the dopant concentration increases and is attributed to the increase in the percentage of the additional phase in the form of microclusters present in the host lattice of Si as evidenced by the change in the bond lengths ΔD (%) from table 7.5.

Table 7.5 Bond lengths obtained from PDF analysis for $Si_{1-x}Ni_x$.

System	I^{st} neighbour(Å)		ΔD (%)	IInd neighbour(Å)		ΔD (%)	IIIrd neighbour(Å)		ΔD (%)
	D_{obs}	D_{cal*}		D_{obs}	D_{cal*}		D_{obs}	D_{cal*}	
Si	2.36	2.30	2.61	3.98	3.92	1.53	5.68	5.76	1.39
$Si_{0.97}Ni_{0.03}$	2.50	2.30	8.69	4.38	3.92	11.73	5.58	5.76	3.13
$Si_{0.94}Ni_{0.06}$	2.18	2.30	5.22	3.86	3.92	1.53	5.70	5.76	1.04
$Si_{0.91}Ni_{0.09}$	2.46	2.30	6.96	4.32	3.92	10.20	6.72	5.76	16.67
$Si_{0.88}Ni_{0.12}$	2.50	2.30	8.69	4.36	3.92	11.22	5.52	5.76	4.17

D_{obs}-Observed nearest neighbor distance
D_{cal*}-Calculated nearest neighbor distance
ΔD-percentage deviation of the nearest neighbor distance
* Using theoretical model (Gretep Software) (Bochu, 2010)

Dilute Magnetic Semiconducting (DMS) Materials, R. Saravanan   Materials Research Forum LLC
Materials Research Foundations **35** (2018)   doi: http://dx.doi.org/10.21741/9781945291777

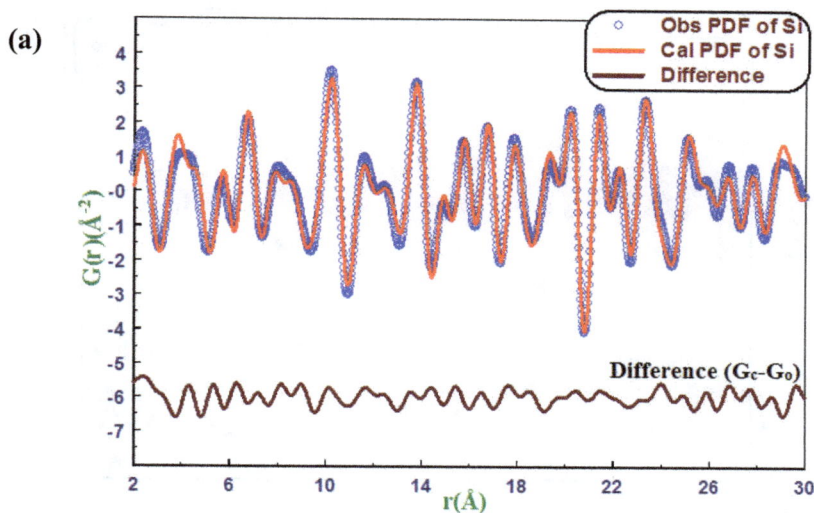

Figure 7.7(a) Fitted pair distribution function (PDF) of Si.

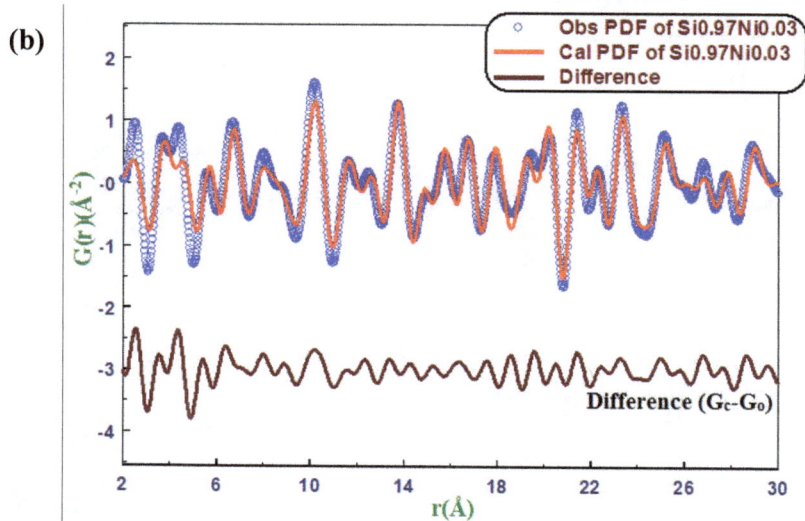

Figure 7.7(b) Fitted pair distribution function (PDF) of $Si_{0.97}Ni_{0.03}$.

(c)

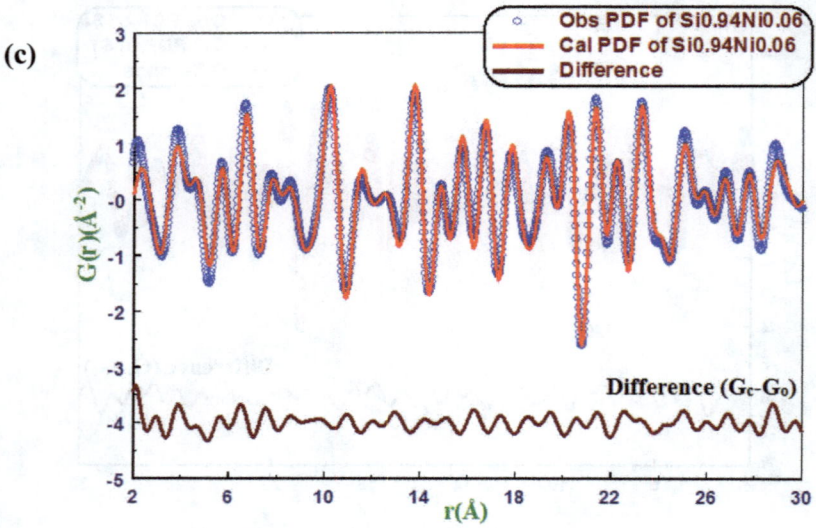

Figure 7.7(c) Fitted pair distribution function (PDF) of $Si_{0.94}Ni_{0.06}$.

(d)

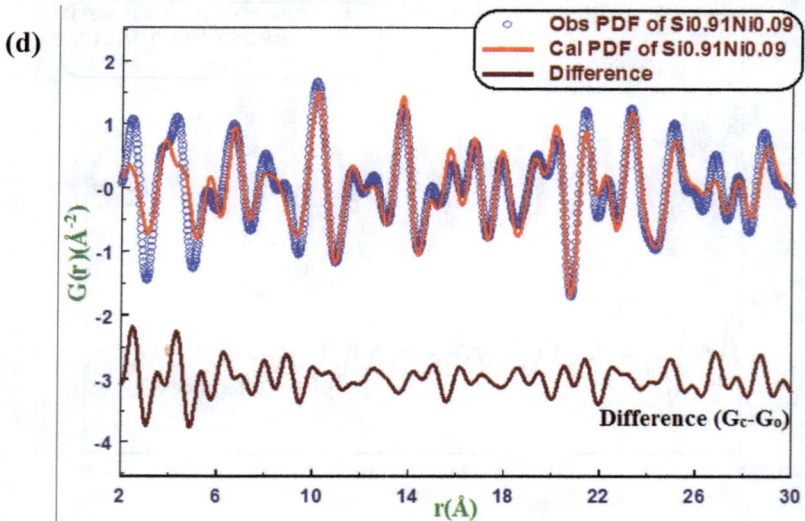

Figure 7.7(d) Fitted pair distribution function (PDF) of $Si_{0.91}Ni_{0.09}$.

(e)

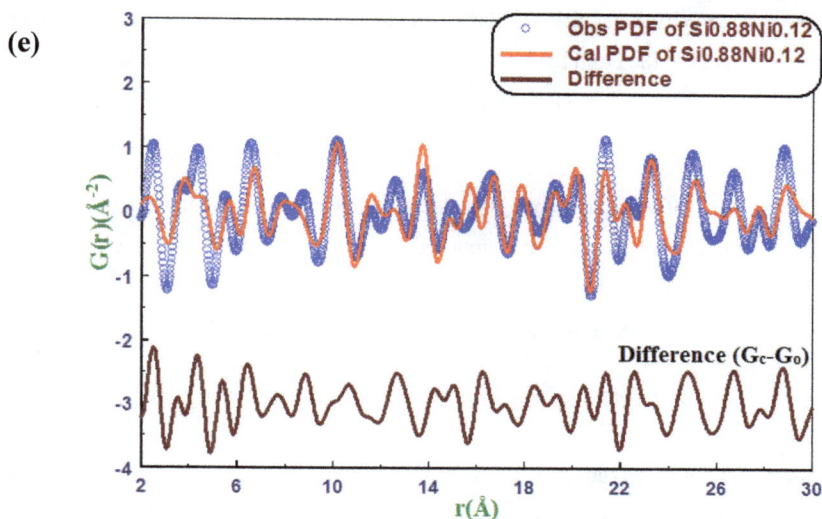

Figure 7.7(e) Fitted pair distribution function (PDF) of $Si_{0.88}Ni_{0.12}$.

7.7 Conclusion

The analysis of the local structure of the prepared DMS samples was done using the pair distribution function (PDF) of the samples evaluated and fitted using the software packages PDFgetX (Jeong et al., 2001) and PDFgui (Farrow et al., 2007) respectively. The nearest neighbor distances of the atoms in each system were estimated and tabulated. The variation in the local structure distribution of the prepared systems with respect to the variation in the dopant concentration was analyzed and the results are presented.

7.7.1 Melt grown $Ge_{1-x}Mn_x$

i.   When the concentration of the dopant increases in the system, the bond lengths of the systems increase.

ii.   The intensity of the pair distribution function decreases with the increase in the dopant concentration evidencing the incorporation of the dopant Mn in the host lattice of Ge.

iii. The existence of the additional phase of Ge_3Mn_5 is found to be the reason behind the mismatch in the bond lengths of the 3rd nearest neighbor and the undulations in the PDFs in the prepared DMS system.

7.7.2 Melt grown $Ge_{1-x}V_x$

i. The presence of vanadium and the dislocation due to the implantation of vanadium atoms in the host Ge matrix are supposed to be the reasons behind the undulations of the PDF below $r = 13$ Å.

ii. The strain introduced in the lattice is reflected through the undulations in the PDFs of the samples.

7.7.3 Melt grown $Ge_{1-x}Co_x$

i. The reduction in the intensity of the peaks with the increase in the concentration of the dopant shows the incorporation of Co atom in the host lattice of Ge.

ii. The difference between the calculated and experimental PDF upto $r = 8$ Å ensures that, GeCo clusters are formed deciding the magnetic behavior of the DMS sample $Ge_{1-x}Co_x$.

7.7.4 Ball milled $Si_{1-x}Mn_x$

i. The intensity of the peaks decreases as the size of the ball milled particles decreases.

ii. As a result of the reduction of the size, there is decrease in the number of atoms at a selected neighbor distance.

iii. The matching of the wriggles in the profiles indicate that all the prepared systems behave in the same way while the amplitude of the function decreases for sample with increased milling time.

7.7.5 Ball milled $Si_{1-x}Ni_x$

i.   The bond length increases due to the expansion of lattice when the bigger Ni atom is incorporated in the regular lattice site of Si when the dopant concentration is $x = 0.03$.

ii. When the dopant concentration increases further, the presence of interstitial Ni microclusters causes the reduction of bond length.

iii. As the dopant concentration increases, the increase in the percentage of the Ni microclusters causes the mismatch in the PDFs of the prepared samples.

References

[1]   Bochu J.L.B., GRETEP, Domaine universitaire BP 46, 38402, 2010, Saint Martin d'Hères <http:/www.inpg.fr/LMGP>.

[2]   Farrow C.L., Juhas P., Liu J.W., Bryndin D., Bozin E.S., Bloch J., Proffen T., Billinge S.J.L., J. Phys.:: Condens. Matter. Phys. 19, 335219 (2007). https://doi.org/10.1088/0953-8984/19/33/335219

[3]   Jeong I.K., Thompson J., Proffen T., Perez A., Billinge S.J.L., J. Appl. Cryst., 34, 536 (2001). https://doi.org/10.1107/S0021889801009207

[4]   Overhauser A.W., J. Phys. Chem. Solids., 13, 71 (1960). https://doi.org/10.1016/0022-3697(60)90128-1

[5]   Proffen T., Billinge S.J.L., Journal of Applied Crystallography, 32, 572 (1999). https://doi.org/10.1107/S0021889899003532

Dilute Magnetic Semiconducting (DMS) Materials, R. Saravanan　　　　Materials Research Forum LLC
Materials Research Foundations **35** (2018)　　　　　doi: http://dx.doi.org/10.21741/9781945291777

Chapter 8

Conclusion

Abstract

Chapter 8 lists down the consolidated discussion of this work.

Keywords

Melt Grown, Ball Milled, Vanadium Doped Germanium, Manganese Doped Germanium, Cobalt Doped Germanium, Manganese Doped Silicon, Nickel Doped Silicon

Contents

An effort has been made to fabricate a series of Ge and Si based diluted magnetic semiconductors and to analyze themfor structural and magnetic properties. The electronic charge density distribution of the fabricated DMS materials is also analyzed. For the fabrication of the desired diluted magnetic semiconductors, two methods have been adopted in this work, *viz.,* melt growth technique and ball milling technique. The results are summarized here.

The materials prepared using melt technique are:

i.　　$Ge_{1-x}Mn_x$ (x = 0, 0.04, 0.06, 0.10)

ii.　　$Ge_{1-x}V_x$ (x = 0.03, 0.06, 0.09)

Dilute Magnetic Semiconducting (DMS) Materials, R. Saravanan   Materials Research Forum LLC
Materials Research Foundations **35** (2018)   doi: http://dx.doi.org/10.21741/9781945291777

iii. $Ge_{1-x}Co_x$ (x = 0.03, 0.06, 0.09)

The materials prepared using ball milling technique are:

  i. $Si_{1-x}Mn_x$ (x = 0.02); milled for 100h and 200h
 ii. $Si_{1-x}Ni_x$ (x = 0, 0.03, 0.06, 0.09, 0.12)

For the characterization of the grown samples, the techniques used are:

  i. Powder X – ray diffraction (Analysis of structural properties)
 ii. Scanning electron microscopy (Morphological structure)
iii. Vibrating sample magnetometry (Analysis of the magnetic properties)

8.1 Melt grown $Ge_{1-x}Mn_x$ (x = 0, 0.04, 0.06, 0.10)

The diffraction peaks from the observed powder X – ray diffractionprofilesof melt grown $Ge_{1-x}Mn_x$have been analyzed and the lattice parameters have been deduced.

The structural analysis of $Ge_{1-x}Mn_x$ reveals:

  i. The doping of Mn in the host Ge matrix has been confirmed from the changes in the lattice parameters.
 ii. An increase in the powder X – ray intensities with Mn concentration confirmsthe doping of Mn in the host Ge matrix.
iii. The contribution of the intermetallic domains to disorder of the lattice has been confirmed by the observed powder X – ray intensity peaks.
 iv. The formation of Mn – poor Ge_8Mn_{11} is preferred when x = 0.04 while formation of both Mn – poor Ge_8Mn_{11} and Mn – rich Ge_3Mn_5 is preferred for higher concentrations of the dopant.
  v. A compression of lattice has been observed when x = 0.04 and expansion of lattice has been observed when x = 0.06. This is due to the hybridization of d orbital of Mn with p orbital of Ge that causes local magnetic moment of Mn atom in the Ge matrix. This magnetic moment increases the mixing of the orbitals and thus affects the structural parameters of $Ge_{1-x}Mn_x$.
 vi. The formation of the intermetallic domains, Ge_3Mn_5 and Ge_8Mn_{11} increases with the dopant concentration and hence, the disorder in the system increases. Therefore, from our observation, x = 0.06 is suggested as the maximum possible dilution for the preparation of DMS $Ge_{1-x}Mn_x$.

The local structural analysis of $Ge_{1-x}Mn_x$ from pair distribution function (PDF) (Proffen et al., 1999) reveals:

i.   The bond – lengths of $Ge_{1-x}Mn_x$ increase with Mn concentration which confirms the incorporation of Mn in the host system.

ii. The undulations in the PDFs confirm the existence of additional phase in $Ge_{1-x}Mn_x$.

iii. The mismatch of bond – lengths at 3rd nearest neighbor evidences the existence of Ge_3Mn_5 in the chosen DMS material.

The analysis of the magnetic hysteresis measurement reveals:

i. $Ge_{1-x}Mn_x$ having the dopant concentrationof $x = 0.04$ behaves as an antiferromagnet due to the domination of the intermetallic domains of Mn – poor Ge_8Mn_{11}.

ii. When $x = 0.06$, $Ge_{1-x}Mn_x$ behaves as a perfect ferromagnet due to the domination of the Mn – rich Ge_3Mn_5 domains.

iii. When $x = 0.10$, $Ge_{1-x}Mn_x$ behaves as a soft ferromagnet.

iv. From all the above observations, it is clear that for the sample to be a perfect ferromagnet, there should be a right stoichiometric composition of Mn and the intermetallic domain Mn – poor Ge_8Mn_{11}.

The charge density analysis reveals:

i.   The charge density around the atomic site of Ge decreases with the increase in the dopant concentration due to the higher covalent radius of Mn than that of Ge.

ii. The mid bond density increases with the dopant concentration in $Ge_{1-x}Mn_x$ due to the formation of intermetallic domains and Andersons mixing effect (Weng et al., 2005) between the orbitals of the host Ge and dopant Mn.

iii. When $x = 0.10$, the value of the interatomic charges is minimum due to the minimum dilution of the dopant Mn in the host lattice.

8.2   Melt grown $Ge_{1-x}V_x$ ($x = 0.03, 0.06, 0.09$)

The powder X – ray diffraction profilesof melt grown $Ge_{1-x}V_x$ have been analyzed. The structural analysis reveals:

i.   A phase pure system of $Ge_{1-x}V_x$ has been prepared and an increase in the total charge in the unit cell (F_{000}) has been observed with the increase in the dopant concentration.

ii.   The observed shifting of the Bragg peaks towards the lower angles with the increase in the dopant concentration is because of the distortion of the Ge host lattice sites by the addition of the dopant.

iii.   An increase in the value of the cell parameterhas been observed with the increase in the dopant concentration.

iv.   The SEM micrographs reveal poor crystallinityfor $x = 0.03$, a better crystalline nature for $x = 0.06$ and a non – uniform grain concentration for $x = 0.09$.

The charge density analysis reveals:

i.   The incorporation of dopant V atoms in the regular lattice site of the host Ge has been confirmed by the increase in the midbond density with the dopant concentration.

ii.   The possible creation of spin – density wave in $Ge_{1-x}V_x$ has been the reason behind the readjustment of conduction electrons resulting in a uniform flat charge density between the atoms.

iii.   When $x = 0.03$, the system exhibits a perfect saddle, and when $x = 0.06$ and 0.09, the charge density is uniform throughout the region and the one – dimensional charge density profile has been uplifted.

iv.   Covalent nature bonding has been observed in the $Ge_{1-x}V_x$ system.

The analysis of the magnetic hysteresis measurements reveals:

i.   The sample with $x = 0.03$ exhibit diamagnetic behavior while the samples having $x = 0.06$ and 0.09 exhibit antiferromagnetic behavior.

ii.   Barkhausen jumps have been observed in the hysteresis loops of $Ge_{1-x}V_x$, which may be due to the crystallographic defects such as dislocations. When $x = 0.09$, the jumps are more, indicating that the dislocations are more.

The local structural analysis of the samples from pair distribution function(PDF) (Proffen et al., 1999)reveals:

i.   The undulations below 13 Å confirm the presence of vanadium in the host lattice of Ge.

ii.   The undulations are also due to the dislocations caused due to the incorporation of the dopant in the host matrix. The strain introduced in the lattice because of the doping process is reflected in the PDF.

8.3   Melt grown $Ge_{1-x}Co_x$ (x = 0.03, 0.06, 0.09)

The observed powder X – ray diffraction profilesof the samples of melt grown $Ge_{1-x}Co_x$ have been analyzed.

The structural analysis of the samples reveals:

  i.   When x = 0.03, the value of the cell parameter increases when compared to that of undoped Ge.

  ii.   When x = 0.06, the value of the cell parameter decreases while it increases for the sample with x = 0.09.

  iii.   The total charge (F_{000}) in the unit cell decreases with increase in the dopant concentration indicating the incorporation of Coin Ge host lattice.

  iv.   Additional peaks have been observed in the diffraction profiles due to the presence of the additional phase GeCo. The intensity of these additional peaks is found to be increasing with the dopant concentration which indicates the inclusion of GeCo clusters.

  v.   It has been observed from the SEM micrographs that,when x = 0.03, $Ge_{1-x}Co_x$ exhibits a nice crystalline nature. A non – uniform grain concentration is observed at x = 0.06. A well pronounced crystalline nature is exhibited in the system when x = 0.09.

The charge density analysis of the systems reveals:

  i.   When x = 0.03, the midbond density in $Ge_{1-x}Co_x$ is higher than that of undoped Ge.

  ii.   An increase in the midbond density has been observed upto the dopant concentration of x = 0.06.

  iii.   A decrease in the midbond density has been observed when x = 0.09, which is an implication that the dopant is not incorporated in the host system properly.

The analysis of the magnetic hysteresis measurements reveals:

  i.   The $Ge_{1-x}Co_x$ system with the dopant concentration of x = 0.03 reveals a diamagnetic behavior.

  ii.   When x = 0.06 and 0.09, the $Ge_{1-x}Co_x$ systemexhibit antiferromagnetic behavior.

  iii.   The observed Barkhausen jumps for the systems are attributed to the addition of GeCo clusters in $Ge_{1-x}Co_x$ system.

The local structural analysis of the samples from pair distribution function(PDF) (Proffen et al., 1999) reveals:

  i.   The intensity peaks of the pair distribution function of the samples decrease with increase in the dopant concentration which evidences the incorporation of the dopant in the host lattice of Ge.
 ii.   The undulations of the profile upto 8Å enable us to understand the fact that the GeCo clusters are formed and affectthemagnetic behavior of $Ge_{1-x}Co_x$system.

8.4   Ball milled $Si_{1-x}Mn_x$ (x = 0.02)

The observed powder X – ray diffraction profilesof the samples of Si and ball milled $Si_{0.98}Mn_{0.02}$have been analyzed and the lattice parameters have been deduced.

The structural analysis of the samples reveals:

  i.   The grain size of the samples decreases due to the process of mechanical alloying. Negligible level of contamination due to the stainless steel container used for ball milling has been observed in the prepared samples.
 ii.   The intensity of the peaks in the X – ray diffraction profiles decreases for the ball milled samples due to the breaking up of the crystalline domains and the increase in amorphous nature.
iii.   An increase in the cell dimension of the ball milled $Si_{0.98}Mn_{0.02}$indicates the incorporation of Mn into the regular lattice site of Si. The volume of the cell increases with the increase in the dopant concentration.
 iv.   The decrease in the Debye – Waller factor (B_{iso}) for $Si_{0.98}Mn_{0.02}$ (100h) indicates the incorporation of the heavier Mn atom at the regular lattice site of Si.
  v.   The increase in the B_{iso} value for $Si_{0.98}Mn_{0.02}$ (200h) is due to the decrease in the size of the grain/domain that accommodates the atoms in the surface of the grain.
 vi.   An additional phase of cubic SiMn with the space group of $P2_13$ and cell value of a_0 = 4.5594Å is found to be present in the ball milled samples and its composition is found to be 2.36%.
vii.   A highly amorphous nature has been exhibited by $Si_{0.98}Mn_{0.02}$ (200h) which makes the refinement procedure difficult and hence not done.

The local structural analysis of the samples from pair distribution function(PDF) (Proffen et al., 1999) reveals:

  i.   The reduction of particle size due to the milling process also results in the reduction in the number of atoms present at a particular distance. This is indicated by the decrease in the relative intensity of the PDF of the samples when the milling time is larger.

ii. As the milling period is increased the number of atoms at a selected neighbor distance is reduced to 57% and then to 4% for the milling period of 100h and 200h respectively.

iii. The matching of the wriggles in the profile shows that all the three samples behave the same way while the amplitude of the function decreases as the milling time is increased.

The charge density analysis of the systems reveals:

i. Undoped Si has been observed to have high midbond density.

ii. Reshaping of the Si atom and an increasein the resistivity of the system $Si_{0.98}Mn_{0.02}$ have been observed with the increase in dopant concentration.

iii. Depletion of charges in the bonding region has been observed in the ball milled $Si_{0.98}Mn_{0.02}$thereby forming closed shell interaction.

iv. When the sample is milled for more than 200h the system can enter into a near insulator phase which is not desired.

The analysis of the magnetic hysteresis measurements reveals:

i. An increase in coercivity and decrease in saturation magnetization with the increase in milling time has been clearly observed from the magnetic hysteresis measurements.

ii. The saturation magnetization of the chosen samples decreases with decreasing particle size.

iii. It can be concluded that the DMS system Si:Mn is ferromagnetic at room temperature when the milling time is suitably chosen.

8.5 Ball milled $Si_{1-x}Ni_x$ (x = 0, 0.03, 0.06, 0.09, 0.12)

The observed powder X – ray diffraction profilesof the samples of ball milled $Si_{1-x}Ni_x$have been analyzed and the lattice parameters have been deduced.

The structural analysis of the samples reveals:

i. The homogeneity of the ball milled $Si_{1-x}Ni_x$ sample has been verified in the SEM measurements. The agglomeration of the particles increases with the increase in the concentration of the dopant in the host lattice.

ii. At x = 0, a phase pure system of $Si_{1-x}Ni_x$with diamond structure and space group $Fd\overline{3}m$ has been observed.

Dilute Magnetic Semiconducting (DMS) Materials, R. Saravanan   Materials Research Forum LLC
Materials Research Foundations **35** (2018)   doi: http://dx.doi.org/10.21741/9781945291777

iii. An additional phase of metallic Ni has been found in a small fraction in all the prepared samples. The Ni phase was found to have FCC structure and the space group $Fm\bar{3}m$.

iv. A shift in the diffraction angles has been observed with increasing dopant concentration in $Si_{1-x}Ni_x$.

v. An increase in the cell dimension has been observed with the increase in the dopant concentration due to the forceful incorporation of the heavier Ni atom in the host lattice of Si.

vi. Additional peaks observed at Bragg angles 44.5° and 52° of the diffraction profiles are attributed to the presence of low soluble Ni clusters. These Ni clusters are found to increase with the increase in the dopant concentration.

vii. The observed shift in the prominent peak (111) is due to the inclusion of Ni in the host lattice of Si.

The local structural analysis of the samples from pair distribution function(PDF) (Proffen et al., 1999) reveals:

i. As the dopant Ni is included in the host lattice Si (at x = 0.03), the bond length increases and this is found to be due to the expansion of lattice happened in order to accommodate the forceful inclusion of the bigger Ni atom in the regular lattice site of Si.

ii. When the dopant concentration increases, the presence of the additional phase of Ni reduces the bond length in $Si_{1-x}Ni_x$ system.

iii. The mismatch between observed and the calculated profiles of pair distribution functions of the samples of $Si_{1-x}Ni_x$ becomes more visible as the dopant concentration increasesdue to the increase in the additional phase.

The analysis of the magnetic hysteresis measurements reveals:

i. All the prepared systems $Si_{1-x}Ni_x$ show room temperature ferromagnetic (RTFM) property which is necessary for a DMS material.

ii. The merging of the hysteresis curves at x = 0.03 and 0.06 shows that the samples enact the same behavior. The ferromagnetic property becomes well pronounced as x increases upto 0.12.

iii. The saturation magnetization for x = 0.03 and x = 0.06 are same whereas it increases threefold when the dopant concentration goes to x = 0.12.

The charge density analysis of the systems reveals:

i. A change in the spherical shape of the charge density of the host Si atoms has been observed with the increase in the dopant concentration.

ii.   The shift in the peak at (111) infers that, the impurity atom prefers to occupy the substitutional position rather than interstitial sites.

iii.   In the ball milled $Si_{1-x}Ni_x$, the intermediate charges between the atoms increases as the dopant Ni is introduced in the host Si and this behavior is well pronounced at 9% and then it enhances to the maximum at 12%.

iv.   The mid bond electron density increases with the increase in the dopant concentration.

8.6   Comparison of the theoretical and experimental charge densities of the Si and Ge based DMS materials

i.   The theoretical electron densities have been calculated for Si based diluted magnetic semiconductors with the dopants Mn, V and Co. From table 8.1 it is clear that, the atomic numbers of the dopants are higher than the host Si. When the concentration of the dopant increases, the attraction to the core increases which leads to the decrease in the value of the mid bond density (Figure 8.1). The values of the theoretical and experimental charge densities of the Si based DMS materials are presented as tables 8.2 and 8.3 respectively.

ii.   The theoretical electron densities were calculated for Ge based diluted magnetic semiconductors with the dopants Mn, V and Co. From table 8.1 it is clear that, the atomic numbers of the dopants are lower that the host Ge. When the concentration of the dopant increases, the value of the mid bond density increases (Figure 8.3). The values of the theoretical and experimental charge densities of Ge based DMS materials are presented as tables 8.4 and 8.5 respectively.

Table 8.1 Atomic number of the hosts and dopants.

Elements	Atomic number
Silicon (Si)	14
Germanium (Ge)	32
Vanadium (V)	23
Manganese (Mn)	25
Cobalt (Co)	27
Nickel (Ni)	28

Table 8.2 Theoretical electron densities of Si based DMS materials.

Composition (x)	Bond length (Å)	Theoretical Electron density (e/Å³)		
		$Si_{1-x}V_x$	$Si_{1-x}Mn_x$	$Si_{1-x}Co_x$
0.02	1.1817	0.2044	0.2035	0.2035
0.04	1.1817	0.2001	0.1980	0.1980
0.06	1.1817	0.1954	0.1921	0.1918
0.08	1.1817	0.1906	0.1858	0.1850
0.10	1.1817	0.1855	0.1791	0.1778

Table 8.3 Experimental electron densities of Si based DMS materials.

System	Bond length (Å)	Experimental charge density (e/Å³)
Si	1.1702	0.2415
$Si_{0.98}Mn_{0.02}$(100h)	1.2025	0.5005
$Si_{0.98}Mn_{0.02}$(200h)	1.1860	0.5173
$Si_{0.97}Ni_{0.03}$	1.1739	0.3193
$Si_{0.94}Ni_{0.06}$	1.1709	0.3690
$Si_{0.91}Ni_{0.09}$	1.1711	0.4087
$Si_{0.88}Ni_{0.12}$	1.1805	0.4678

Table 8.4 Theoretical electron densities of Ge based DMS materials.

Composition (x)	Bond length (Å)	Theoretical Electron density (e/Å³)		
		$Ge_{1-x}V_x$	$Ge_{1-x}Mn_x$	$Ge_{1-x}Co_x$
0.02	1.2353	0.5485	0.5444	0.5414
0.04	1.2353	0.5614	0.5541	0.5476
0.06	1.2353	0.5744	0.5633	0.5538
0.08	1.2353	0.5876	0.5726	0.5599
0.10	1.2353	0.6009	0.5814	0.5661

Dilute Magnetic Semiconducting (DMS) Materials, R. Saravanan   Materials Research Forum LLC
Materials Research Foundations **35** (2018)   doi: http://dx.doi.org/10.21741/9781945291777

Table 8.5 Experimental electron densities of Ge based DMS materials.

System	Bond length (Å)	Experimental charge density (e/Å³)
$Ge_{0.96}Mn_{0.04}$	1.2180	0.3222
$Ge_{0.94}Mn_{0.06}$	1.2230	0.3400
$Ge_{0.90}Mn_{0.10}$	1.2222	0.3170
$Ge_{0.97}V_{0.03}$	1.2189	0.3217
$Ge_{0.94}V_{0.06}$	1.2184	0.4358
$Ge_{0.91}V_{0.09}$	1.2180	0.4551
$Ge_{0.97}Co_{0.03}$	1.2308	0.3580
$Ge_{0.94}Co_{0.06}$	1.2307	0.4246
$Ge_{0.91}Co_{0.09}$	1.2300	0.4156

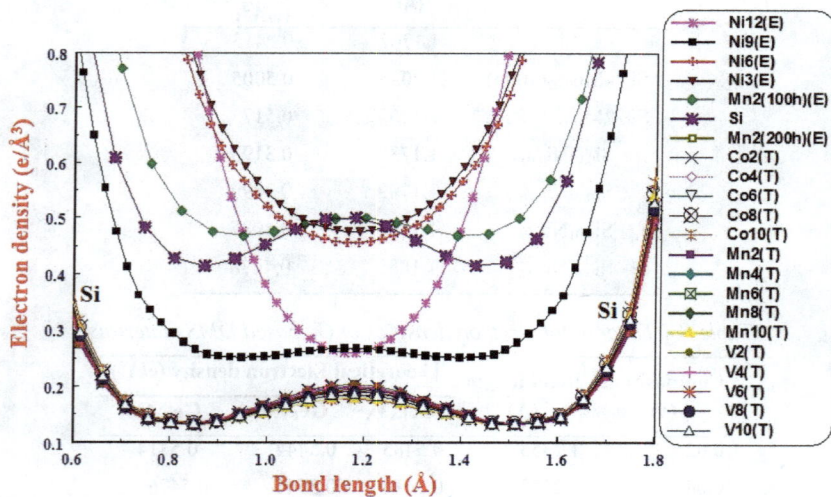

Figure 8.1 Comparison of the experimental and theoretical electron densities of Si based DMS materials (E – Experimental electron density; T – Theoretical electron density).

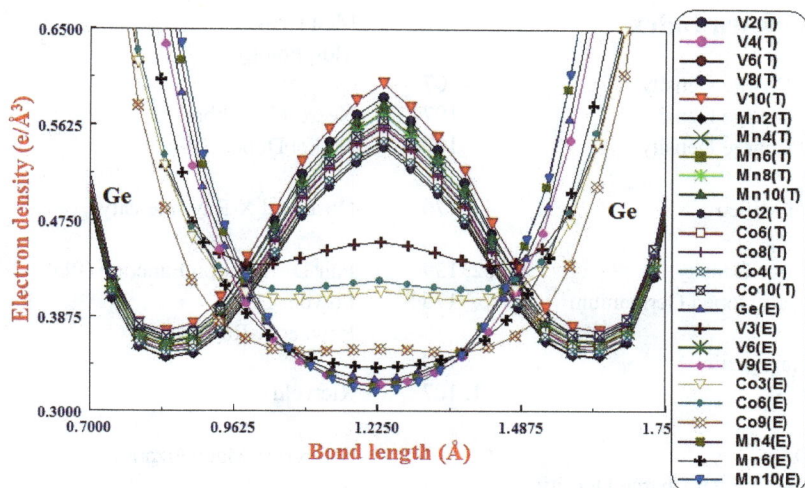

Figure 8.2 Comparison of the experimental and theoretical electron densities of Ge based DMS materials (E – Experimental electron density; T – Theoretical electron density).

To conclude, it can be mentioned here that the detailed electronic structure and properties can be evolved using the best possible experimental tool like X-ray diffraction and mathematical tools like maximum entropy method (Collins, 1982). The same kind of work on other technologically important materials will throw light on these materials and their suitability for the technology that uses electronics.

References

[1]   Collins D.M., Nature, 298, 49 (1982). https://doi.org/10.1038/298049a0

[2]   Proffen T., Billinge S.J.L., Journal of Applied Crystallography,32, 572(1999). https://doi.org/10.1107/S0021889899003532

[3]   Weng H., Dong J., Physical Review B., 71, 035201 (2005). https://doi.org/10.1103/PhysRevB.71.035201

Keyword Index

About the Author

Dr Ramachandran Saravanan, has been associated with the Department of Physics, The Madura College, affiliated with the Madurai Kamaraj University, Madurai, Tamil Nadu, India from the year 2000. He is the head of the Research Centre and PG department of Physics. He worked as a research associate during 1998 at the Institute of Materials Research, Tohoku University, Sendai, Japan and then as a visiting researcher at Centre for Interdisciplinary Research, Tohoku University, Sendai, Japan up to 2000.

Earlier, he was awarded the Senior Research Fellowship by CSIR, New Delhi, India, during Mar. 1991 - Feb.1993; awarded Research Associateship by CSIR, New Delhi, during 1994 – 1997. Then, he was awarded a Research Associateship again by CSIR, New Delhi, during 1997- 1998. Later he was awarded the Matsumae International Foundation Fellowship in1998 (Japan) for doing research at a Japanese Research Institute (not availed by him due to the simultaneous occurrence of other Japanese employment).

He has guided eleven Ph.D. scholars as of 2017, and about five researchers are working under his guidance on various research topics in materials science, crystallography and condensed matter physics. He has published around 140 research articles in reputed Journals, mostly International, apart from around 50 presentations in conferences, seminars and symposia. He has also guided around 60 M.Phil. scholars and an equal number of PG students for their projects. He has attracted government funding in India, in the form of Research Projects. He has completed two CSIR (Council of Scientific and Industrial Research, Govt. of India), one UGC (University Grants Commission, India) and one DRDO (Defense Research and Development Organization, India) research projects successfully and is proposing various projects to Government funding agencies like CSIR, UGC and DST.

He has written 8 books in the form of research monographs including; "Experimental Charge Density - Semiconductors, oxides and fluorides" (ISBN-13: 978-3-8383-8816-8; ISBN-10:3-8383-8816-X), "Experimental Charge Density - Dilute Magnetic Semiconducting (DMS) materials" (ISBN-13: 978-3-8383-9666-8; ISBN-10: 3-8383-9666-9) and "Metal and Alloy Bonding - An Experimental Analysis" (ISBN -13: 978-1-4471-2203-6). He has committed to write several books in the near future.

His expertise includes various experimental activities in crystal growth, materials science, crystallographic, condensed matter physics techniques and tools as in slow evaporation, gel, high temperature melt growth, Bridgman methods, CZ Growth, high vacuum sealing etc. He and his group are familiar with various equipment such as: different types of cameras; Laue, oscillation, powder, precession cameras; Manual 4-circle X-ray diffractometer, Rigaku 4-circle automatic single crystal diffractometer, AFC-5R and AFC-7R automatic single crystal diffractometers, CAD-4 automatic single crystal diffractometer, crystal pulling instruments, and other crystallographic, material science related instruments. He and his group have sound computational capabilities on different types of computers such as: IBM – PC, Cyber180/830A – Mainframe, SX-4 Supercomputing system – Mainframe. He is familiar with various kind of software related to crystallography and materials science. He has written many computer software programs himself as well. Around twenty of his programs (both DOS and GUI versions) have been included in the SINCRIS software database of the International Union of Crystallography.